ANTI-TECH REVOLUTION: WHY AND HOW

ANTI-TECH REVOLUTION: WHY AND HOW

THEODORE JOHN KACZYNSKI

Second Edition

FITCH & MADISON
PUBLISHERS

Second edition, 2020.

Published by Fitch & Madison Publishers.

Inquiries to the publisher should be addressed to Fitch & Madison Publishers, 15150 North Hayden Road, Suite 210, Scottsdale, AZ 85260, Tel: 602-457-4800, Fax: 602-457-4802, or via e-mail at info@fitchmadison.com.

Fitch & Madison and Fitch & Madison Publishers are registered trademarks of Fitch & Madison Publishers, LLC, an Arizona limited liability company.

www.fitchmadison.com

Theodore John Kaczynski does not receive any remuneration for this book.

First Fitch & Madison edition published 2016.

Printed in the United States of America

∞ This paper meets the requirements of ANSI/NISO Z39.48-1992 (Permanence of Paper).

10 9 8 7 6 5 4 3

Publisher's Cataloging-in-Publication Data

Kaczynski, Theodore John, 1942- author.
Anti-tech revolution : why and how / Theodore John Kaczynski. — 2nd edition.
pages cm
Includes bibliographical references and index.
LCCN 201992978
ISBN 978-1-944228-02-6

1. Technology and civilization. 2. Technology—Social aspects. 3. Revolutions—History
4. Science—Philosophy & Social aspects.
5. Environmental degradation. 6. Social action. I. Title.

T14.5.K325 2019 303.48'3—dc23

Were there but an Adam and an Eve left in every continent, and left free, it would be better than it now is.

— Thomas Jefferson

CONTENTS

The epigraph is from Jefferson's letter to William Short, January 1793, quoted by David McCullough, *John Adams*, Simon & Schuster, New York, 2002, p. 438.

PREFACE TO THE FIRST EDITION

I. There are many people today who see that modern society is heading toward disaster in one form or another, and who moreover recognize technology as the common thread linking the principal dangers that hang over us.[1] Nearly all such people fall into one of two categories:

First, there are those who are appalled at what technology is doing to our society and our planet, but are not motivated to take any action against the technological system because they feel helpless to accomplish anything in that direction. They read an anti-tech book—say, for example, Jacques Ellul's *Technological Society*—and it makes them feel better because they've found someone who has eloquently articulated their own anxieties about technology. But the effect soon wears off and their discomfort with the technological world begins to nag them again, so they turn for relief to another anti-tech book—Ivan Illich, Kirkpatrick Sale, Daniel Quinn, my own *Industrial Society and Its Future*, or something else—and the cycle repeats itself. In other words, for these people anti-tech literature is merely a kind of therapy: It alleviates their discomfort with technology, but it does not serve them as a call to action.

In the second category are people who are appalled at modern technology and actually aspire to accomplish something against the technological system, but have no practical sense of how to go about it. At a purely *tactical* level some of these people may have excellent practical sense; they may know very well, for example, how to organize a demonstration against some particular atrocity that is being committed against our environment. But when it comes to grand strategy[2] they are at a loss. Most perhaps recognize that any victory against an environmental atrocity or other technology-related evil can only be temporary, at best, as long as the technological system remains in existence. But they can think of nothing better to do than to continue attacking particular evils while vaguely hoping that their work will somehow help to solve the overall problem of technology. In reality their work is counterproductive, because it distracts attention from the technological system itself as the underlying source of the evils and leads people to focus instead on problems of limited significance that moreover cannot be permanently solved while the technological system continues to exist.

The purpose of this book is to show people how to begin thinking in practical, grand-strategic terms about what must be done in order to get our society off the road to destruction that it is now on.

On the basis of past experience I feel safe in saying that virtually all people—even people of exceptional intelligence—who merely read this book once or twice at an ordinary pace will miss many of its most important points. This book, therefore, is not a book to be *read*; it is a book to be *studied* with the same care that one would use in studying, for example, a textbook of engineering. There is of course a difference between this book and a textbook of engineering. An engineering textbook provides precise rules which, if followed mechanically, will consistently give the expected results. But no such precise and reliable rules are possible in the social sciences. The ideas in this book therefore need to be applied thoughtfully and creatively, not mechanically or rigidly. Intelligent application of the ideas will be greatly facilitated by a broad knowledge of history and some understanding of how societies develop and change.

II. This book represents only a part, though the most important part, of a longer work that I hope to publish later. I've been anxious to get the most important part of the work into print as soon as possible, because the growth of technology and the destruction of our environment move at an ever-accelerating rate, and the time to begin organizing for action is—as soon as possible. Moreover, I'm 72 years old, and I could be put out of action at any time by some medical misfortune, so I want to get the most important material into print while I can.

The entire work—the part published here together with the parts that at present exist only in the form of imperfect drafts—goes far beyond my earlier works, *Industrial Society and Its Future* and *Technological Slavery*, and it represents the more-or-less final result of a lifetime of thought and reading—during the last thirty-five years, intensive thought and specifically purposeful reading. The factual basis of the work is drawn primarily from my reading over all those years, and especially from the reading I've done since 1998 while confined in a federal prison. As of 2011, however, there remained important loose ends that needed to be tied up, gaps that needed to be filled in, and I've been able to tie up those loose ends and fill in those gaps only with the generous help of several people outside the prison who have delved for the information I've requested and have answered almost all of the questions—sometimes very difficult questions—that I've asked them.

My thanks are owing above all to Susan Gale. Susan has played the key role in this project and has been indispensable. She has been my star researcher, producing more results and solving more problems, by far, than anyone else; she has ably coordinated the work of other researchers and has done most of the typing.

After Susan, the most important person in this project has been Dr. Julie Ault. Julie has read drafts of the various chapters and has called my attention to many weak points in the exposition. I've tried to correct these, though I haven't been able to correct all of them to my (or, I assume, her) satisfaction. In addition, Julie has provided valuable advice on manuscript preparation.[3] But most important of all has been the encouragement I've taken from the fact of having an intellectual heavyweight like Julie Ault on my side.

Several people other than Susan have made important research contributions through steady work over a period of time: Brandon Manwell, Deborah, G.G. Gómez, Valerie v.E., and one other person whose name will not be mentioned here. Patrick S. and another person, who prefers not to be named, have provided critically important financial support and have been helpful in other ways as well.

The foregoing are the people who have made major contributions to the project, but I owe thanks also to nine other people whose contributions have been of lesser magnitude: Blake Janssen, Jon H., and Philip R. each dug up several pieces of information for me; Lydia Eccles, Dr. David Skrbina, Isumatag (pseudonym), and Último Reducto (pseudonym) have called my attention to information or sent me copies of articles that I've found useful; Lydia has also performed other services, and an assistant of Dr. Skrbina's typed early drafts of Chapter Three and Appendix Three. On the legal front, I owe thanks to two attorneys for their pro-bono assistance: Nancy J. Flint, who took care of copyright registration, and Edward T. Ramey, whose intervention removed a bureaucratic obstacle to the preparation of this book.

My thanks to all!

III. Despite the generous help I've received, I've had to make use at many points of sources of information that are of doubtful reliability; for example, media reports (all too often irresponsible!) or encyclopedia articles, which, because of their necessary brevity, commonly give only sketchy accounts of the subjects they cover. None of the individuals named above are in any way responsible for the resulting defects of this book. It is only since 2011 that I've had people who have been willing and able to spend substantial amounts of time and effort in doing research for me, and all of them have had to carry on simultaneously with other necessary aspects of their lives, such as earning a living. If I had asked them to find solid authority for every piece of information for which I've relied on a questionable source, the completion of this book would have been delayed for a matter of years. I do not believe that my use of questionable sources of information will be found to weaken significantly the arguments or the conclusions that I offer in this book. Even if some of the bits of information I've cited turn out to be false, inaccurate, or misleading, the basic structure of the book will remain sound.

IV. *Note on referencing*. In the notes that follow each chapter or appendix, I generally cite sources of information by giving the author's last name and a page number. The reader can find the author's full name, the title of the book or article cited, the date of publication, and other necessary information by looking up the author's name in the List of Works Cited that appears at the end of the book. When a source without named author is cited, the reader will in some cases be able to find additional information about the source by consulting the list of works without named author that concludes the List of Works Cited.

Two abbreviations are used repeatedly in the notes:

"ISAIF" refers to my *Industrial Society and Its Future*, of which only one correct version has been published in English; it appears on pages 36–120 of my book *Technological Slavery* (Feral House, 2010).

"NEB" means *The New Encyclopaedia Britannica*, Fifteenth Edition. The Fifteenth Edition has been modified repeatedly, so "NEB" is always followed by a date in parentheses that indicates the particular version of NEB that is cited. For example, "NEB (2003)" means the version of *The New Encyclopaedia Britannica* that bears the copyright date 2003.

Ted Kaczynski
May 2014

NOTES

1. I've received many letters from such people, not only from within the United States but from a score of countries around the world.

2. "Tactics," "strategy," and "grand strategy" are, in origin at least, military terms. Tactics are techniques used for the immediate purpose of winning a particular battle; strategy deals with broader issues and longer intervals of time, and includes advance preparations for winning a battle or a series of battles; grand strategy addresses the entire process of achieving a nation's objectives through warfare, and takes into account not only the strictly military aspect of the process but also the political, psychological, economic, etc. factors involved. See, e.g., NEB (2003), Vol. 29, "War, Theory and Conduct of," p. 647. The terms "tactics," "strategy," and "grand strategy" are used by analogy in contexts that have nothing to do with warfare or the military.

3. For reasons connected with the need to get the manuscript for the present work prepared quickly, I've disregarded some of Julie Ault's recommendations concerning manuscript preparation. Needless to say, Julie is in no way responsible for any resulting defects that may be found in this book.

PREFACE TO THE SECOND EDITION

I. This second edition of *Anti-Tech Revolution* closely follows the format of the first edition, but has been improved in a variety of minor ways—as by citation of better authority for some statements of fact and correction of a few errors of fact—and also in four ways that I consider moderately important: 1. In Part III of Chapter One it is argued that the power of revolutionary dictators like Hitler and Stalin was far from truly absolute. In the first edition that argument was weak, at least in regard to Hitler; it has now been greatly strengthened. 2. In Part II of Chapter Three, the discussion of early Christianity was weak because based on insufficient information. Since then I've been able to read two excellent books on early Christianity, and I've strengthened the discussion accordingly. 3. In Part III of Chapter Three I've strengthened somewhat the discussion of Rule (iv) (that a revolutionary movement should strive to exclude unsuitable persons who may seek to join it), which in the first edition was weaker than it is now. 4. Also in Part III of Chapter Three, I've brought the discussion of Mexican politics up to date (meaning up to July 2018). In some other—though minor—ways as well the present edition has been brought up to date, but no systematic effort has been made to update every part of the book. The amount of research and other work required would have been prohibitive, and in any case this book is not designed to be a current-events report; its purpose is to elucidate certain general principles, which in most instances can be done just as well with old facts as with fresh ones.

II. In Part II of the preface to the first edition I wrote: "This book represents only a part, though the most important part, of a longer work that I hope to publish later." Since then I've decided that the additional material I had hoped to include in the longer work can best be placed in the second volume of the revised edition of my book *Technological Slavery*. In view of my age and the difficult circumstances under which I have to work, it remains an open question whether I will ever be able to complete that second volume.

To some people the present work may seem nihilistic, inasmuch as it focuses on the need to eliminate modern technology but says very little about *positive* values opposed to those of the technological system.

Elsewhere, however, I've discussed the positive value of wild nature and of a life lived close to nature,[1] and if I live to complete the second volume of *Technological Slavery* it will include a section titled "Is There Such a Thing as Wilderness? Is There Such a Thing as the Balance of Nature?," which should do a great deal to dispel any aura of nihilism that may seem to surround radical opposition to the technological system.

III. What was said in Part III of the preface to the first edition concerning my use of doubtful sources of information—for example, media reports—applies equally to this second edition. Moreover, as pointed out in note 102 to Chapter Three, the need for brevity has forced me in some cases to compress my account of historical events to the point of serious oversimplification. For these reasons the present work is not suitable for use as a source-book of facts. Readers who require a high degree of reliability and factual accuracy should consult the sources I've cited, evaluate them, and then conduct further research as needed.

Readers who are familiar with some of the works I've referenced in the notes may wonder why, in some cases, I've cited facts from a given book but failed to cite from the same book other facts that would have been relevant to my arguments. This is explained by the difficult circumstances in which I'm placed. Prisoners here are not allowed to accumulate many books in their cells; consequently I read a book, take notes on it, then send it to friends on the outside. But in taking notes I can't fully anticipate my future needs for information, so I've often failed to record information that I would have found relevant a few years later. A similar problem occurs with books from the prison library; I read them and take notes on them, but in many cases the books subsequently disappear from the library—because they've been damaged or stolen by prisoners, or have simply worn out, or for some other reason.

IV. I could not have prepared this second edition of *Anti-Tech Revolution* without the help of several people outside the prison. Susan Gale, above all, has continued to play a central and indispensable role in my writing projects. She is the best and most important researcher, she coordinates the work of other researchers, she is an excellent typist, and she helps in a variety of ways too numerous to mention.

Others who have helped with research are Traci J. Macnamara, Elizabeth Tobier, T.F., N.P., and S.T. Elizabeth has been especially generous in ordering books for me at her own expense; Dr. Susie Meister, L.R.F., T.F., C.H., and S.T. too have ordered books for me at their own expense, while Lydia Eccles and Manuel Monteiro have sent me valuable articles from periodicals or from the Internet. Patrick S. has provided important financial support. I owe all of these people a debt of gratitude, and I especially want to thank Manuel for having arranged the publication in Europe of the first edition of *Anti-Tech Revolution*.

V. The note on referencing that concludes the preface to the first edition applies equally to the present edition. I only need to add that references like "Kaczynski, Letter to David Skrbina (+ date)," "Kaczynski, 'The System's Neatest Trick,'" and so forth, direct the reader to parts of my book *Technological Slavery*.

Ted Kaczynski
October 2018

NOTE

1. See mainly: ISAIF, ¶¶ 183–84, 197–99. Kaczynski, Letters to David Skrbina: Aug. 29, 2004; Sept. 18, 2004, point (ii); Oct. 12, 2004, Part II; Nov. 23, 2004, Parts III.D&E. In the 2010 Feral House edition of Kaczynski, the *Blackfoot Valley Dispatch* interviews, pp. 394–407.

CHAPTER ONE

The Development of a Society Can Never Be Subject to Rational Human Control

> Adonde un bien se concierta
> hay un mal que lo desvía;
> mas el bien viene y no acierta,
> y el mal acierta y porfía.
>
> — Diego Hurtado de Mendoza (1503–1575)[1]

> The wider the scope of my reflection on the present and the past, the more am I impressed by their mockery of human plans in every transaction.
>
> — Tacitus[2]

I. In specific contexts in which abundant empirical evidence is available, fairly reliable short-term prediction and control of a society's behavior may be possible. For example, economists can predict some of the immediate consequences for a modern industrial society of a rise or a fall in the interest rates. Hence, by raising or lowering interest rates they can manipulate such variables as the levels of inflation and of unemployment.[3] Indirect consequences are harder to predict, and prediction of the consequences of more elaborate financial manipulations is largely guesswork. That's why the economic policies of the U.S. government are subject to so much controversy: No one knows for certain what the consequences of those policies really are.

Outside of contexts in which abundant empirical evidence is available, or when longer-term effects are at issue, successful prediction—and therefore successful management of a society's development—is far more difficult. In fact, failure is the norm.

- During the first half of the second century BC, sumptuary laws (laws intended to limit conspicuous consumption) were enacted in an effort to forestall the incipient decadence of Roman society. As is usual with sumptuary laws, these failed to have the desired effect, and the decay of Roman mores continued unchecked.[4] By the early first century BC, Rome had become politically unstable. With the help of soldiers under his command, Lucius Cornelius Sulla seized control of the city, physically exterminated the opposition, and carried out a comprehensive program of reform that was intended to restore stable government. But Sulla's intervention only made the situation worse, because he had killed off the "defenders of lawful government" and had filled the Senate with unscrupulous men "whose tradition was the opposite of that sense of mission and public service that had animated the best of the aristocracy."[5] Consequently the Roman political system continued to unravel, and by the middle of the first century BC Rome's traditional republican government was essentially defunct.
- In Italy during the 9th century AD certain kings promulgated laws intended to limit the oppression and exploitation of peasants by the aristocracy. "The laws proved futile, however, and aristocratic landowning and political dominance continued to grow."[6]
- Simón Bolívar was the principal leader of the revolutions through which Spain's American colonies achieved their independence. He had hoped and expected to establish stable and "enlightened" government throughout Spanish America, but he made so little progress toward that objective that he wrote in bitterness shortly before his death in 1830: "He who serves a revolution plows the sea." Bolívar went on to predict that Spanish America would "infallibly fall into the hands of the unrestrained multitude to pass afterward to those of... petty tyrants of all races and colors... [We will be] devoured by all crimes and extinguished by ferocity [so that] the Europeans will not deign to conquer us... ."[7] Allowing for a good deal of exaggeration attributable to the emotion under which Bolívar wrote, this prediction held (roughly) true for a century and a half after his death. But notice that Bolívar did not arrive at this prediction until too late; and that it was a very general prediction that asserted nothing specific.
- In the United States during the late 19th century there were

> worker-housing projects sponsored by a number of individual philanthropists and housing reformers. Their objective was to show that efforts to

> improve the living conditions of workers could be combined with... profits of 5 percent annually. ...
>
> Reformers believed that the model dwellings would set a standard that other landlords would be forced to meet... mostly because of the workings of competition. Unfortunately, this solution to the housing problem did not take hold.... The great mass of urban workers... were crowded into... tenements that operated solely for profit.[8]

It is not apparent that there has been any progress over the centuries in the capacity of humans to guide the development of their societies. Relatively recent (post-1950) efforts in this direction may seem superficially to be more sophisticated than those of earlier times, but they do not appear to be more successful.

- The social reform programs of the mid-1960s in the United States, spearheaded by President Lyndon Johnson, revealed that beliefs about the causes and cures of such social problems as crime, drug abuse, poverty, and slums had little validity. For example, according to one disappointed reformer:

> Once upon a time we thought that if we could only get our problem families out of those dreadful slums, then papa would stop taking dope, mama would stop chasing around, and junior would stop carrying a knife. Well, we've got them in a nice new apartment with modern kitchens and a recreation center. And they're the same bunch of bastards they always were.[9]

This doesn't mean that all of the reform programs were total failures, but the general level of success was so low as to indicate that the reformers did not understand the workings of society well enough to know what should be done to solve the social problems that they addressed. Where they achieved some modest level of success they probably did so mainly through luck.[10]

- It was once believed that the "emergence of a truly interconnected world" via the Internet would be "a step toward cross-cultural cooperation and global enlightenment. As societies communicate more freely, ... empathy will be nourished, the truth will be easier to find, and many causes of conflict will wither. ... The age of social media, in other words, should be an age of peace and understanding."[11]

The actual result has been nothing of the kind. Instead, the Internet has played a major role in the development of what many people call a "post-truth" or "post-fact" society—a society in which it becomes ever more difficult to escape systematic deception or ascertain the objective truth.[12] The Internet also serves as a deadly tool for terrorists, and as a weapon for unscrupulous national leaders who intentionally promote conflict.[13]

One could go on and on citing examples like the foregoing ones. One could also cite many examples of efforts to control the development of societies in which the immediate goals of the efforts have been achieved. But in such cases the longer-term consequences for society as a whole have not been what the reformers or revolutionaries have expected or desired.[14]

- The legislation of the Athenian statesman Solon (6th century BC) was intended to abolish hektemorage (roughly equivalent to serfdom) in Attica while allowing the aristocracy to retain most of its wealth and privilege. In this respect the legislation was successful. But it also had unexpected consequences that Solon surely would not have approved. The liberation of the "serfs" resulted in a labor shortage that led the Athenians to purchase or capture numerous slaves from outside Attica, so that Athens was transformed into a slave society. Another indirect consequence of Solon's legislation was the Peisistratid "tyranny" (populist dictatorship) that ruled Athens during a substantial part of the 6th century BC.[15]

- Otto von Bismarck, one of the most brilliant statesmen in European history, had an impressive list of successes to his credit. Among other things:

—He achieved the unification of Germany in 1867–1871.

—He engineered the Franco-Prussian war of 1870–71, but his successful efforts for peace thereafter earned him the respect of European leaders.

—He successfully promoted the industrialization of Germany.

—By such means he won for the monarchy the support of the middle class.

—Thus Bismarck achieved his most important objective: He prevented (temporarily) the democratization of Germany.

—Though Bismarck was forced to resign in 1890, the political structure he had established for Germany lasted until 1918, when it was brought down by the German defeat in World War I.[16]

Notwithstanding his remarkable successes Bismarck felt that he had failed, and in 1898 he died an embittered old man.[17] Clearly, Germany was not going the way he had intended. Probably it was the resumption of Germany's slow drift toward democratization that angered him most.

But his bitterness would have been deeper if he had foreseen the future. One can only speculate as to what the history of Germany might have been after 1890 if Bismarck hadn't led the country up to that date, but it is certain that he did not succeed in putting Germany on a course leading to results of which he would have approved; for Bismarck would have been horrified by the disastrous war of 1914–18, by Germany's defeat in it, and above all by the subsequent rise of Adolf Hitler.

- In the United States, reformers' zeal led to the enactment in 1919 of "Prohibition" (prohibition of the manufacture, sale, or transportation of alcoholic beverages) as a constitutional amendment. In terms of its immediate objectives Prohibition was rather successful, for it reduced per capita consumption of alcohol in the United States by some sixty or seventy percent, it diminished the incidence of alcohol-related diseases and deaths, and it "eradicated the saloon." On the other hand it provided criminal gangs with opportunities to make huge profits through the smuggling and/or the illicit manufacture of alcoholic drinks; thus Prohibition greatly promoted the growth of organized crime. In addition, it led to the corruption both of public institutions and of individual citizens. It became clear that Prohibition was a serious mistake, and it was repealed through another constitutional amendment in 1933.[18]

- The so-called "Green Revolution" of the latter part of the 20th century—the introduction of new farming technologies and of recently developed, highly productive varieties of grain—was supposed to alleviate hunger in the Third World by providing more abundant harvests. It did indeed provide more abundant harvests. But: "[A]lthough the 'Green Revolution' seems to have been a success as far as the national total cereal production figures are concerned, a look at it from the perspective of communities and individual humans indicates that the problems have far outweighed the successes... ."[19] In some parts of the world the consequences of the Green Revolution have been nothing short of catastrophic. For example, in the Punjab (a region lying partly in India and partly in Pakistan), the Green Revolution has ruined "thousands of hectares of [formerly] productive land," and has led to severe lowering of the water table, contamination of the water with pesticides and fertilizers, numerous cases of cancer (probably due to the contaminated water), and many suicides. " 'The green revolution has brought us only downfall,' says Jarnail Singh... . 'It ruined our soil, our environment, our water table. Used to be we had fairs in villages where people would come together and have fun.

Now we gather in medical centers.'"[20]

From other parts of the world as well come reports of negative consequences, of varying degrees of severity, that have followed the Green Revolution. These consequences include economic, behavioral, and medical effects in addition to environmental damage (e.g., desertification).[21]

• In 1953, U.S. President Eisenhower announced an "Atoms for Peace" program according to which the nations of the world were supposed to pool nuclear information and materials under the auspices of an international agency. In 1957 the International Atomic Energy Agency was established to promote the peaceful uses of atomic energy, and in 1968 the United Nations General Assembly approved a "non-proliferation" treaty under which signatories agreed not to develop nuclear weapons and in return were given nuclear technology that they were supposed to use only for peaceful purposes.[22] The people involved in this effort should have known enough history to realize that nations generally abide by treaties only as long as they consider it in their own (usually short-term) interest to do so, which commonly is not very long. But apparently the assumption was that the nations receiving nuclear technology would be so grateful, and so happy cooperating in its peaceful application, that they would forever put aside the aspirations for power and the bitter rivalries that throughout history had led to the development of increasingly destructive weapons.

This idea seems to have originated with scientists like Robert Oppenheimer and Niels Bohr who had helped to create the first atomic bomb.[23] That physicists would come up with something so naïve was only to be expected, since specialists in the physical sciences almost always are grossly obtuse about human affairs. It seems surprising, however, that experienced politicians would act upon such an idea. But then, politicians often do things for propaganda purposes and not because they really believe in them.

The "Atoms for Peace" idea worked fine—for a while. Some 140 nations signed the non-proliferation treaty in 1968 (others later),[24] and nuclear technology was spread around the world. Iran, in the early 1970s, was one of the countries that received nuclear technology from the U.S.[25] And the nations receiving such technology didn't try to use it to develop nuclear weapons. Not *immediately*, anyway. Of course, we know what has happened since then. "[H]ard-nosed politicians and diplomats [e.g., Henry Kissinger]...argue that proliferation of nuclear weapons is fast approaching a 'tipping point' beyond which it will be impossible to check their spread."

These "veterans of America's cold-war security establishment with impeccable credentials as believers in nuclear deterrence" now claim that such weapons "ha[ve] become a source of intolerable risk."[26] And there is the inconvenient fact that the problem of safe disposal of radioactive waste from the *peaceful* uses of nuclear energy still has not been solved.[27]

The "Atoms for Peace" fiasco suggests that humans' capacity to control the development of their societies not only has failed to progress, but has actually retrogressed. Neither Solon nor Bismarck would have supported anything as stupid as "Atoms for Peace."

II. There are good reasons why humans' capacity to control the development of their societies has failed to progress. In order to control the development of a society you would have to be able to predict how the society would react to any given action you might take, and such predictions have generally proven to be highly unreliable. Human societies are complex systems—technologically advanced societies are most decidedly complex—and prediction of the behavior of complex systems presents difficulties that are not contingent on the present state of our knowledge or our level of technological development.

> [U]nintended consequences [are] a well-known problem with the design and use of technology... . The cause of many [unintended consequences] seems clear: The systems involved are complex, involving interaction among and feedback between many parts. Any changes to such a system will cascade in ways that are difficult to predict; this is especially true when human actions are involved.[28]

Problems in economics can give us some idea of how impossibly difficult it would be to predict or control the behavior of a system as complex as a modern human society. It is convincingly argued that a modern economy can never be rationally planned to maximize efficiency, because the task of carrying out such planning would be too overwhelmingly complex.[29] Calculation of a rational system of prices for the U.S. economy alone would require manipulation of a conservatively estimated 6×10^{13} (sixty trillion!) simultaneous equations.[30] That takes into account only the economic factors involved in establishing prices and leaves out the innumerable psychological, sociological, political, etc., factors that continuously interact with the economy.

Even if we make the wildly improbable assumption that the behavior of our society could be predicted through the manipulation of, say, a million trillion simultaneous equations and that sufficient computing power to conduct such manipulation were available, collection of the data necessary for insertion of the appropriate numbers into the equations would be impracticable,[31] especially since the data would have to meet impossibly high standards of precision if the predictions were expected to remain valid over any considerable interval of time. Edward Lorenz, a meteorologist, was the first to call widespread attention to the fact that even the most minute inaccuracy in the data provided can totally invalidate a prediction about the behavior of a complex system. This fact came to be called the "butterfly effect" because in 1972, at a meeting of the American Association for the Advancement of Science, Lorenz gave a talk that he titled "Predictability: Does the Flap of a Butterfly's Wings in Brazil Set Off a Tornado in Texas?"[32] Lorenz's work is said to have been the inspiration for the development of what is called "chaos theory"[33]—the butterfly effect being an example of "chaotic" behavior.

Chaotic behavior is not limited to complex systems; in fact, some surprisingly simple systems can behave chaotically.[34] The *Encyclopaedia Britannica* illustrates this with a purely mathematical example. Let A and x_0 be any two given numbers with $0<A<4$ and $0<x_0<1$, and let a sequence of numbers be generated according to the formula $x_{n+1}=Ax_n(1-x_n)$. For certain values of A, e.g., A=3.7, the sequence behaves chaotically: In order to bring about a *linear* increase in the number of terms of the sequence that one can predict to a reasonable approximation, one needs to achieve an *exponential* improvement in the accuracy of one's estimate of x_0. In other words, in order to predict the nth term of the sequence, one needs to know the value of x_0 with an error not exceeding 10^{-kn}, k a constant.[35] This is characteristic of chaotic systems generally: Any small extension of the range of prediction requires an exponential improvement in the accuracy of the data.

> [A]ll chaotic systems share the property that every extra place of decimals in one's knowledge of the starting point only pushes the horizon [of predictability] a small distance away. In practical terms, the horizon of predictability is an impassable barrier. ... [O]nce it becomes clear how many systems are sufficiently nonlinear to be considered for chaos, it has to be recognized that prediction may be limited to short stretches set by the

> horizon of predictability. Full comprehension... must frequently remain a tentative process... with frequent recourse to observation and experiment in the event that prediction and reality have diverged too far.[36]

It should be noted that the Heisenberg Uncertainty Principle sets an absolute limit to the precision of data used for the prediction of physical phenomena. This principle, which implies that certain events involving subatomic particles are unpredictable, is inferred mathematically from other known laws of physics; hence, successful prediction at the subatomic level would entail violations of the laws of physics. If a prediction about the behavior of a macroscopic system requires data so precise that their accuracy can be disturbed by events at the subatomic level, then no reliable prediction is possible. Hence, for a chaotic physical system, there is a point beyond which the horizon of predictability can never be extended.

Of course, the behavior of a human society is not in every respect chaotic; there are empirically observable historical trends that can last for centuries or millennia. But it is wildly improbable that a modern technological society could be free of all chaotic subsystems whose behavior is capable of affecting the society as a whole, so it is safe to assume that the development of a modern society is necessarily chaotic in at least some respects and therefore unpredictable.

This doesn't mean that no predictions at all are possible. In reference to weather forecasting the *Britannica* writes:

> It is highly probable that atmospheric movements... are in a state of chaos. If so, there can be little hope of extending indefinitely the range of weather forecasting except in the most general terms. There are clearly certain features of climate, such as annual cycles of temperature and rainfall, which are exempt from the ravages of chaos. Other large-scale processes may still allow long-range prediction, but the more detail one asks for in a forecast, the sooner it will lose its validity.[37]

Much the same can be said of the behavior of human society (though human society is far more complex even than the weather). In some contexts, reasonably reliable and specific short-term predictions can be made, as we noted above in reference to the relationship between interest rates, inflation, and unemployment. Long-term predictions of an imprecise and nonspecific character are often possible; we've already mentioned

Bolívar's correct prediction of the failure of stable and "enlightened" government in Spanish America. (Here it is well to note that predictions that something will *not* work can generally be made with greater confidence than predictions that something *will* work.[38]) But reliable long-term predictions that are at all specific can seldom be made.

There are exceptions. Moore's Law makes a specific prediction about the rate of growth of computing power, and as of 2012 the law has held true for some fifty years.[39] But Moore's Law is not an inference derived from an understanding of society, it is simply a description of an empirically observed trend, and no one knows how long the trend will continue. The law may have predictable consequences for many areas of technology, but no one knows in any specific way how all this technology will interact with society as a whole. Though Moore's Law and other empirically observed trends may play a useful role in attempts to foresee the future, it remains true that any effort to understand the development of our society must (to borrow the *Britannica's* phrases) "remain a tentative process... with frequent recourse to observation and experiment. . . ."

But just in case someone declines to assume that our society includes any important chaotic components, let's suppose for the sake of argument that the development of society could in principle be predicted through the solution of some stupendous system of simultaneous equations and that the necessary numerical data at the required level of precision could actually be collected. No one will claim that the computing power required to solve such a system of equations is currently available. But let's assume that the unimaginably vast computing power predicted by Ray Kurzweil[40] will become a reality for some future society, and let's suppose that such a quantity of computing power would be capable of handling the enormous complexity of the *present* society and predicting its development over some substantial interval of time. It does not follow that a future society of that kind would have sufficient computing power to predict its *own* development, for such a society necessarily would be incomparably more complex than the present one: The complexity of a society will grow right along with its computing power, because the society's computational devices are part of the society.

There are in fact certain paradoxes involved in the notion of a system that predicts its own behavior. These are reminiscent of Russell's Paradox in set theory[41] and of the paradoxes that arise when one allows a statement to talk about itself (e.g., consider the statement, "This statement is false").

When a system makes a prediction about its own behavior, that prediction may itself change the behavior of the system, and the change in the behavior of the system may invalidate the prediction. Of course, not every statement that talks about itself is paradoxical. For example, the statement, "This statement is in the English language" makes perfectly good sense. Similarly, many predictions that a system may make about itself will not be self-invalidating; they may even cause the system to behave in such a way as to fulfill the prediction.[42] But it is too much to hope for that a society's predictions about itself will *never* be (unexpectedly) self-invalidating.

A society's ability to predict its own behavior moreover would seem to require something like complete self-knowledge, and here too one runs into paradoxes. We need not discuss these here; some thought should suffice to convince the reader that any attempt to envision a system having complete self-knowledge will encounter difficulties.

Thus, from several points of view—past and present experience, complexity, chaos theory, and logical difficulties (paradoxes)—it is clear that no society can accurately predict its own behavior over any considerable span of time. Consequently, no society can be consistently successful in planning its own future in the long term.

This conclusion is in no way unusual, surprising, or original. Astute observers of history have known for a long time that a society can't plan its own future. Thus Thurston writes: "[N]o government has ever been able physically to manage the total existence of a country, ...or to foresee all the complications that would ensue from a decision made at the center."[43]

According to Henry Kissinger: "History is a tale of efforts that failed, of aspirations that weren't realized, of wishes that were fulfilled and then turned out to be different from what one expected."[44]

Norbert Elias wrote: "[T]he actual course of... historical change as a whole is intended and planned by no-one."[45] And: "Civilization... is set in motion blindly, and kept in motion by the autonomous dynamics of a web of relationships... ."[46]

III. The expected answer to the foregoing will be: Even granting that the behavior of a society is unpredictable in the long term, it may nevertheless be possible to steer a society rationally by means of continual short-term interventions. To take an analogy, if we let a car without a driver roll down a rugged, irregular hillside, the only prediction we can make is that the car will not follow any predetermined course but will bounce around erratically. However, if the car has a driver, he may be able to steer it so as to avoid the worst bumps and make it roll instead through relatively smooth places. With a good deal of luck he may even be able to make the car arrive approximately at a preselected point at the foot of the hill. For these purposes the driver only needs to be able to predict very roughly how far the car will veer to the right or to the left when he turns the steering wheel. If the car veers too far or not far enough, he can correct with another turn of the wheel.

Perhaps something similar could be done with an entire society. It is conceivable that a combination of empirical studies with increasingly sophisticated theory may eventually make possible fairly reliable short-term predictions of the way a society will react to any given change—just as fairly reliable short-term weather forecasting has become possible. Perhaps, then, a society might be successfully steered by means of frequent, intelligent interventions in such a way that undesirable outcomes could usually be avoided and some desirable outcomes achieved. The steering process would not have to be infallible; errors could be corrected through further interventions. Just possibly, one might even hope to succeed in steering a society so that it would arrive in the long run at something approximating one's conception of a good society.

But this proposal too runs into difficulties of a fundamental kind. The first problem is: Who decides what outcomes are desirable or undesirable, or what kind of "good" society should be our long-term goal? There is never anything resembling general agreement on the answers to such questions. Friedrich Engels wrote in 1890:

> History is made in such a way that the final result always arises from the conflicts among many individual wills, each of which is made into what it is by a multitude of special conditions of life; thus there are innumerable intersecting forces, an infinite collection of parallelograms of forces, and from them emerges a resultant—the historical event—which from another point of view can be regarded as the product of one power that, as a whole, operates unconsciously and without volition. For what each individual wants runs up against the opposition of every other, and what comes out of it all is something that no one wanted.[47]

Norbert Elias, who was not a Marxist, made a very similar remark:

> [F]rom the interweaving of countless individual interests and intentions—whether tending in the same direction or in divergent and hostile directions—something comes into being that was planned and intended by none of these individuals, yet has emerged nevertheless from their intentions and actions.[48]

Even in those rare cases in which almost everyone agrees on a policy, effective implementation of the policy may be prevented by what is called the "problem of the commons." The problem of the commons consists in the fact that it may be to everyone's advantage that everyone should act in a certain way, yet it may be to the advantage of each individual to act in a contrary way.[49] For example, in modern society it is to everyone's advantage that everyone should pay a portion of his income to support the functions of government. Yet it is to the advantage of each individual to keep all his income for himself, and that's why hardly anyone pays taxes voluntarily, or pays more than he has to.

The answer to the foregoing arguments will be that political institutions exist precisely in order to resolve such problems: The concrete decisions made in the process of governing a society are not the resultant of conflicts among the innumerable individual wills of the population at large; instead, a small number of political leaders are formally empowered (through elections or otherwise) to make necessary decisions for everyone, and to enact laws that compensate for the problem of the commons by compelling individuals to do what is required for the common welfare (for example, laws that compel payment of taxes). Since the top political leaders are relatively few in number, it is not unreasonable to hope that they can resolve their differences well enough to steer the development of a society rationally.

Actually, experience shows that when the top political leaders number more than, say, half a dozen or so, it must seriously be doubted whether they can ever resolve their differences well enough to be able to govern in a consistently rational way. But even where no conflicts exist among the top leaders, the real power of such leaders is very much less than the power that is formally assigned to them. Consequently, their ability to steer the development of their society rationally is extremely limited at best.

When this writer was in the Sacramento County Main Jail in 1996–98, he had some interesting conversations with the jail administrator,

Lieutenant Dan Lewis. In the course of one such conversation, on December 31, 1996, Lewis complained that it was not easy to get some of his officers to follow his orders, and he described the problems that a person in a position of formal power faces when he tries to exert that power to make his organization do what he wants it to do. If the leader takes measures that are resented by too many of the people under his command, he will meet with so much resistance that his organization will be paralyzed.[50]

It's not only jail administrators whose power is far more limited than it appears to an outsider. Julius Caesar reportedly said, "The higher our station, the less is our freedom of action."[51] According to an English author of the 17th century: "Men in great place (saith one) are thrice servants; servants of the sovereign, or state; servants of fame; and servants of business. So as they have no freedom, neither in their persons, nor in their actions, nor in their times."[52] U.S. President Abraham Lincoln wrote: "I claim not to have controlled events, but confess plainly that events have controlled me."[53]

While F.W. de Klerk was President of South Africa, Nelson Mandela asked him why he did not prevent acts of violence that in some cases were being carried out with the collusion of the police. De Klerk replied, "Mr. Mandela, when you join me [as a member of the government] you will realise I do not have the power which you think I have."[54] It's possible that de Klerk was pleading powerlessness as an excuse for tolerating violence that in reality he might have been able to prevent. Nevertheless, when Mandela himself became President, he "quickly realized, as de Klerk had warned him, that a President had less power than he appeared to. He could rule effectively only through his colleagues and civil servants, who had to be patiently persuaded. . . ."[55]

In line with this, a thorough student of the American presidency, Clinton Rossiter, has explained how severely the power of the President of the United States is limited, not only by public opinion and by the power of Congress, but also by conflicts with members of his own administration who, in theory, are totally under his command.[56] Rossiter refers to "the trials undergone by [Presidents] Truman and Eisenhower in persuading certain chiefs of staff, whose official lives depend entirely on the President's pleasure, to shape their acts and speeches to the policies of the administration."[57] One of our most powerful presidents, Franklin D. Roosevelt, complained:

> The Treasury is so large and far-flung and ingrained in its practices that I find it is almost impossible to get the actions and results I want.... But the Treasury is not to be compared with the State Department. You should go through the experience of trying to get any changes in the thinking, policy and action of the career diplomats and then you'd know what a real problem was. But the Treasury and the State Department put together are nothing compared with the Na-a-vy. The admirals are really something to cope with—and I should know. To change anything in the Na-a-vy is like punching a feather bed. You punch it with your right and you punch it with your left until you are finally exhausted, and then you find the damn bed just as it was before you started punching.[58]

Roosevelt's capable successor in the presidency, Harry S. Truman, said:

> [P]eople talk about the powers of a President, all the powers that a Chief Executive has, and what he can do. Let me tell you something—from experience!
>
> The President may have a great many powers given to him by the Constitution and may have certain powers under certain laws which are given to him by the Congress of the United States; but the principal power that the President has is to bring people in and try to persuade them to do what they ought to do without persuasion. That's what I spend most of my time doing. That's what the powers of the President amount to.[59]

Thus, concentration of formal power in the hands of a few top leaders by no means liberates decision-making from Engels's "conflicts among many individual wills." Some people may be surprised to learn that this is true even in a society governed by a single, theoretically absolute ruler.

• From 200 BC to 1911 AD, all Chinese dynasties were headed by an emperor who "was the state's sole legislator, ultimate executive authority, and highest judge. His pronouncements were, quite literally, the law, and he alone was not bound by his own laws."[60] The emperor was supposed to be restrained by "Confucian norms and the values perpetuated by the scholar-official elite,"[61] but in the absence of an explicit codification or any mechanism for enforcement, these restraints were effective against the emperor only to the extent that some of his subjects were brave enough to challenge him on their own initiative, though the emperor, "if he insisted, would prevail."[62]

More important, therefore, were the practical limitations to which the emperor was subject. "As the head of a vast governmental apparatus... he was... forced to delegate his powers to others who conducted the routine operations of government.... Institutions inherited from previous dynasties were the main vehicles through which he delegated political responsibilities," for "in seeking alternatives to that immediate past, one had no models outside of China to draw upon."[63] Needless to say, the actual power wielded by an emperor depended on the energy and ability of the individual who occupied the office at any given time,[64] but it seems clear that that power was in every case far less than what might naively be inferred from the fact that the emperor's word was law.

To illustrate the practical limitations on the emperor's power with a concrete example, in 1069 AD the emperor Shenzong (Shen-tsung), having recognized the brilliance of the political thinker Wang Anshi (An-shih), appointed him Vice Chief Councillor in charge of administration and gave him full power to implement his ideas in the emperor's name.[65] Wang based his reforms on thorough study, but both he and the emperor failed to take account of the bitter opposition that the new policies would arouse among those whose private interests were threatened by them.[66] "Even in the short run, the cost of the divisive factionalism that the reforms generated had disastrous effects."[67] Opposition to Wang was so intense that he resigned permanently in 1076, and during the eight years following Shenzong's death in 1085 most of the reforms were rescinded or drastically revised.[68] Under two subsequent emperors, Zhezong (Che-tsung; reigned in effect, circa 1093–1100) and Huizong (Hui-tsung; reigned 1100–1126), some of the reforms were restored, but "Wang's own former associates were gone, and his policies became nothing more than an instrument in bitter political warfare."[69] "[A]lthough Emperor Huizong's reign saw some of the reform measures reinstated, the atmosphere at his court was not one of high-minded commitment,"[70] but was characterized by "debased political behavior."[71] "Leading officials engaged in corrupt practices," and the rapacity of the emperor's agents "aroused serious revolts of people who in desperation took up arms against them."[72] The fall of the Northern Song (Sung) Dynasty in 1126–27 marked the final demise of whatever was left of Wang's reforms.[73]

- Norbert Elias makes clear that the "absolute" monarchs of the "Age of Absolutism" in Europe were not so absolute as they seemed.[74] For example, Louis XIV of France is generally seen as the archetype of the "absolute" monarch; he could probably have had any individual's head chopped off at will. But by no means could he use his power freely:

> The vast human network that Louis XIV ruled ha[d] its own momentum and its own centre of gravity which he had to respect. It cost immense effort and self-control to preserve the balance of people and groups and, by playing on the tensions, to steer the whole.[75]

Elias might have added that Louis XIV could "steer" his realm only within certain narrow limits. Elias himself refers elsewhere to "the realization that even the most absolute government is helpless in the face of the dynamisms of social development... ."[76]

• The theoretically absolute emperor Joseph II ruled Austria from 1780 to 1790 and instituted major reforms of a "progressive" (i.e., modernizing) character. But:

"By 1787 resistance to Joseph and his government was intensifying. ...Resistance simmered in the Austrian Netherlands... .

"[By 1789]... The war [against the Turks] caused an outpouring of popular agitation against his foreign policy, the people of the Austrian Netherlands rose in outright revolution, and reports of trouble in Galicia increased. ...

"Faced with these difficulties, Joseph revoked many of the reforms that he had enacted earlier. ...

"...[Joseph II] tried to do too much too quickly and so died a deeply disappointed man."[77]

Especially to be noted is the fact that Joseph II failed even though most of his reforms were modernizing ones; that is, they merely attempted to accelerate Austria's movement in obedience to a powerful pre-existing trend in European history.

Revolutionary dictators of the 20th century, such as Hitler and Stalin, were probably more powerful than traditional "absolute" monarchs, because the revolutionary character of their regimes had done away with many of the traditional, formal or informal social structures and customary restraints that had curbed the "legitimate" monarchs' exercise of their power.[78] But even the revolutionary dictators' power was in practice far less than absolute.

• In the Soviet Union between 1934 and 1941, the Stalin regime was unable to regulate its own labor force, for the "demand for labor created a situation that overrode… the efforts of the regime to control labor through legislation."[79] The government naturally wanted a stable work-force consisting of workers who would remain at their jobs as long as they were needed, but in practice workers "continued to change jobs at a high rate."[80] Laws were evaded or simply ignored, and "hardly slowed down the movement of workers."[81]

More significantly, the Terror of the middle to late 1930s was not a calculated and effective measure undertaken by Stalin to crush resistance to his rule. Instead, a frightened dictator initiated a process that rapidly spiraled out of his control. "Stalin was a man initiating and reacting to developments, not the cold mastermind of a plot to subdue the party and the nation." "It now appears that Stalin and his close associates, having helped create a tense and ugly atmosphere, nonetheless repeatedly reacted [during the Terror] to events they had not planned or foreseen." "An atmosphere of panic had set in reminiscent of the European witch-hunts… ." "Stalin seems to have become steadily more worried as the purges uncovered alleged spies and Trotskyites. Finally he struck at them, almost incoherently. [¶] During 1937 and 1938 events spun out of… control." "[T]he police fabricated cases, tortured people not targeted in Stalin's directives, and became a power unto themselves." "Terror was producing avoidance of responsibility, which was dysfunctional. Whatever the goal at the top, events were again out of control." "[Stalin] reacted, and over-reacted, to events. … He was sitting at the peak of a pyramid of lies and incomplete information… ." "The evidence is now strong that [Stalin] did not plan the Terror."[82]

One of the consequences of the Terror was the elimination of almost all of the trained and experienced officers from the higher ranks of the Soviet army and navy, with the result that Stalin's military machine was crippled.[83] This was at least part of the reason for the Soviet collapse before the German onslaught in 1941.

- During the 1930s, when the Hitler regime was rearming Germany in preparation for anticipated warfare, resistance by the working class "kept the government from curtailing the production of consumers' goods, although civilian output interfered seriously with arms production."[84] In 1936, "a kind of popular uprising in Münsterland" forced the Nazis to replace the crucifixes that they had removed from school buildings, and there were other instances in which fear of resistance from the churches led the regime to moderate its policies.[85] It can hardly be open to doubt that study of the inner workings of the Third Reich would reveal many additional ways in which the Nazis' policies were limited by the anticipated public reaction to them.

Ulam notes that Hitler's original choice for commander-in-chief of the German army was vetoed by his officers "on the grounds that the man was too much of a Nazi," and he implies that Hitler had far less control over the German military establishment than Stalin had over the Soviet one.[86] But Hitler surely would have been able to eliminate all resistance from his officer corps by conducting a thorough purge like that carried out by Stalin. Actually, Hitler was well aware that there was serious disaffection among his generals,[87] yet, until the assassination attempt of July 20, 1944, he never undertook a major purge of the German military.[88] Why didn't he?

The answer to that question helps to show how limited are the options available even to an "absolute" ruler. A dictator in the position of a Hitler or a Stalin has essentially only two choices: He can carry out a thorough purge of his officer corps, as Stalin did, in which case he cripples his military machine by eliminating most of its trained and experienced leaders; or he can leave his officer corps largely intact, as Hitler did, in which case he risks being overthrown by his own generals. Hitler was bent on military conquest, for which he needed an efficient army, so he gambled on retaining his trained and experienced officers even though he knew that many of them were opposed to his policies and some even aspired to remove him from power. Hitler won his gamble in the sense that he remained in power until defeated militarily by the Allies, but he did so only through an astonishing series of lucky breaks. Rothfels[89] marvels at Hitler's incredible luck: Again and again, from 1938 up to July 20, 1944, the Führer's own officers tried to assassinate him or carry out a coup against him, but he was always saved at the last moment by some chance circumstance.[90]

It's worth noting that, in Ulam's opinion, Stalin too needed "fantastic luck" in order to gain power and retain it as long as he did.[91]

Quite apart from any resistance by subordinates or other "conflicts among individual wills" within a system, purely technical factors narrowly limit the options open even to a leader whose power over his system is theoretically absolute.

- In Frank Norris's immortal novel, *The Octopus*—about wheat farmers whose livelihood is destroyed by railroad rate increases—the protagonist, Presley, confronts the apparently ruthless businessman

Shelgrim, President of the railroad. But Shelgrim tells him:

" 'You are dealing with forces, young man, when you speak of wheat and the railroads, not with men. ... Men have only little to do with the whole business. ... Blame conditions, not men.'

" 'But—but', faltered Presley, 'You are the head, you control the road.'

" '...Control the road! ... I can go into bankruptcy if you like. But otherwise, if I run my road as a business proposition, I can do nothing. I can *not* control it.' "[92]

The Octopus is a work of fiction, but it does truthfully represent, in dramatized form, the economic realities of the era in which Norris wrote (about the end of the 19th and the beginning of the 20th century). At that time, "railway labor and material costs" had increased, and "many American railroads, already struggling to stay alive economically, could not afford rate reductions." State railroad commissions "seeking... ways of establishing fair, 'scientific' rates" found that "there was no such thing as 'scientific' rate making. They discovered that it was extraordinarily difficult to define the 'public interest' or to take the rate question 'out of politics.' Setting rates meant assigning economic priorities, and someone—shipper, carrier, consumer—inevitably got hurt."[93] So it's likely that a railroad like Shelgrim's would indeed have gone bankrupt if it had tried to set rates in such a way as to treat everyone "fairly" and humanely.

It is probably true in general that the ruthless behavior of business enterprises is more often compelled by economic realities than voluntarily chosen by a rapacious management.

• In the 1830s, at an early stage of the U.S. industrial revolution, the textile manufacturers of Massachusetts treated their employees benevolently. Nowadays their system would no doubt be decried as "paternalistic," but in material terms the workers could consider themselves fortunate, for working conditions and housing were very good by the standards of the time. But during the 1840s the situation of the workers began to deteriorate. Wages were reduced, hours of work increased, and greater effort was demanded of the workers; and this was the result not of employers' greed but of market conditions that grew out of economic competition.[94] "As business became nationwide... the competition of different manufacturing areas meant that prices and wages were no longer determined by local conditions. They fluctuated as a consequence of economic changes wholly beyond the control of the employers or workers immediately concerned."[95]

• A recent (2012) article by Adam Davidson discusses some of the reasons behind the problem of unemployment in the U.S. Taking as an example a company that he has personally investigated, Davidson writes: "It's tempting to look to the owners of Standard Motor Products and ask them to help [unskilled workers]: to cut costs a little less relentlessly, take slightly lower profits, and maybe even help solve America's jobs crisis in some small way." Davidson then goes on to explain why a company like Standard Motor Products would be unable to survive in the face of competition if it did not cut costs relentlessly and, therefore, replace human workers with machines whenever it was profitable to do so.[96] Here again we see that "[t]he businessman... [is] only the agent of economic forces and developments beyond his control."[97]

In the last two examples the options open to leaders of organizations were limited not by technical factors alone, but by these in conjunction with competition from outside the organization. But even independently of external competition and of any "conflict of wills" within a system, technical factors by themselves severely limit the choices available to the system's leaders. Not even dictators can escape these limitations.

• In the *Encyclopaedia Britannica* article on Spain we find: "For almost 20 years after the [Spanish Civil War], the [Franco regime] followed a policy of... national economic self-sufficiency.... Spain's policies of economic self-sufficiency were a failure, and by the late 1950s the country was on the verge of economic collapse."[98]

Unwilling to rely solely on the foregoing brief passage for twenty years of Spanish economic history, this writer consulted a Spanish correspondent, who sent him copies of pages from relevant historical works.[99] It turned out that the *Britannica's* account—perhaps unavoidably in view of its brevity—was oversimplified to the point of being seriously misleading. Among other things, it isn't clear to what extent Spain's policy of self-sufficiency was voluntarily chosen and to what extent it was forced on the country, first by the conditions prevailing during World War II and later by the Western democracies' hostility to the authoritarian regime of Franco. Much of this history is beyond the understanding of those of us who have no specialized knowledge of economics, but one thing does emerge clearly: Quite apart from any external competition or internal conflict, economic reality imposes narrow limits on what even an authoritarian regime can do with a nation's economy. A dictator cannot run an economy the way a general runs an army—by giving orders from above—because the economy won't follow orders.[100] In other words, not even a powerful dictator like Francisco Franco can overrule the laws of economics.

Nor can idealistic zeal overcome those laws.

• In the years following the Cuban Revolution of 1956–59, U.S. media propaganda portrayed Fidel Castro as motivated by a lust for power, but actually Castro started out with generalized humanitarian and democratic goals.[101] Once he had overthrown the Batista government, he found that, despite the immense power conferred on him by his personal charisma,[102] the options open to him were extremely limited. Circumstances forced him to choose between democracy and the deep social reforms that he envisioned; he couldn't have both. Since his basic goals were his social ones he had to abandon democracy, become a dictator, and Stalinize and militarize Cuban society.[103]

There can be no doubt about the idealistic zeal of the Cuban revolutionaries,[104] and Castro was as powerful as any charismatic dictator could ever be.[105] Even so, the revolutionary regime was unable to control the development of Cuban society: Castro admitted that he had failed to curb the bureaucratic tendencies of Cuba's administrative apparatus.[106] Notwithstanding the regime's strong ideological opposition to racism, "the drive to promote... blacks and mixed race Cubans to leadership positions within the government and Party" was only partly successful, as Castro himself acknowledged.[107] In fact, Cuban efforts to combat racism do not seem to have been any more successful than those of the United States.[108] The Castro regime achieved no more than minimal success in its attempt to free the Cuban economy from its almost total dependence on sugar and to industrialize the country.[109] To survive at all economically, the regime was forced to abandon its attempt to build "socialism" (as conceived by Cuba's idealistic leaders) within a short period. It was found necessary instead to make ideologically painful compromises with economic reality,[110] and even with these compromises the Cuban economy has remained no more than barely viable.[111]

A contributing factor in Cuba's economic failure was the embargo imposed by the United States: U.S. firms were forbidden to trade with Cuba. But this factor was not decisive, and not as important as admirers of the Castro regime liked to think. Cuba could trade with most of the economically important countries of the world other than the U.S., and was even able to trade indirectly with major U.S. corporations by dealing with their subsidiaries in other countries.[112] The embargo was far less important than Cuba's inability to free itself from its excessive dependence on sugar or even to run its sugar industry efficiently.[113] Another factor in Cuba's economic failure was a lack of cooperation within Cuban society—Engels's "conflicts among many individual wills." There were absenteeism, passive resistance to production quotas, and "stolid peasant resistance."[114] "Individualistic" tendencies led to pilfering, waste, and even to major criminal activity.[115] In addition, there were conflicts within the Cuban power-structure.[116] Almost certainly, however, the decisive factor in Cuba's failure has been the Castro regime's refusal to comply with the technical requirements for economic success: The regime compromised its ideology only as far as was necessary for bare survival, and declined to accept those elements of the free market and of capitalism that might have made vigorous development possible. That this factor was decisive is shown by the fact that purely socialist economies have failed all over the world.[117]

IV. There is yet another—and critically important—reason why a society cannot "steer" itself in the manner suggested at the beginning of Part III of this chapter: Every complex, large-scale society is subject to internal developments generated by "natural selection" operating on systems that exist within the society. This factor is discussed at length in Chapter Two; here we will only sketch the argument in the briefest possible terms.

Through a process analogous to biological evolution there arise, within any complex, large-scale society, self-preserving or self-reproducing systems large and small (including, for example, business enterprises, political parties or movements, open or covert social networks such as networks of corrupt officials) that struggle to survive and propagate themselves. Because power is a cardinal tool for survival, these systems compete for power.

Biological organisms, evolving through natural selection, eventually invade every niche in which biological survival is possible at all, and, whatever measures may be taken to suppress them, some organisms will find ways of surviving nonetheless. Within any complex, large-scale society, a similar process will produce self-propagating systems that will invade every corner and circumvent all attempts to suppress them. These systems will compete for power without regard to the objectives of any government (or other entity) that may try to steer the society. Our argument—admittedly impossible at present to prove conclusively—is that these self-propagating systems will constitute uncontrollable forces that will render futile in the long run all efforts to steer the society rationally. For details, see Chapter Two.

V. Notwithstanding all the arguments we've reviewed in the present chapter up to this point, let's make the unrealistic assumption that techniques for manipulating the internal dynamics of a society will some day be developed to such a degree that a single, all-powerful leader (we'll be charitable and call him a philosopher-king[118] rather than a dictator)—or a group of leaders small enough (< 6?) to be free of "conflicts among individual wills" within the group—will be able to steer a society as suggested at the beginning of Part III, above.

The notion of authoritarian rule by a single leader or a small group of leaders is not as far-fetched as it may appear to the citizens of modern liberal democracies. Many people in the world already live under the authority of one man or a few, and when the technological society gets itself into sufficiently serious trouble, as it is likely to do in the coming decades, even the citizens of liberal democracies will begin looking for solutions that today seem out of the question. During the Great Depression of the 1930s, many Americans—mainstream people, not kooks out on the fringes—felt disillusioned with democracy[119] and advocated rule by a dictator or an oligarchy (a "supercouncil" or a "directorate").[120] Many admired Mussolini.[121] During the same period, many Britons admired Hitler's Germany. "Lloyd George's reaction to Hitler was typical: 'If only we had a man of his supreme quality in England today,' he said."[122]

Returning, then, to our hypothetical dictator, or philosopher-king as we've decided to call him, we'll assume, however implausibly, that he will somehow be able to overcome the problems of complexity, of the conflicts of many individual wills, of resistance by subordinates, and of the competitive, power-seeking groups or systems that will evolve within any complex, large-scale society. Even under this unreal assumption we will still run into fundamental difficulties.

The first problem is: Who is going to choose the philosopher-king and how will they put him into power? Given the vast disparities of goals and values ("conflicts among individual wills") in any large-scale society, it is hardly likely that the rule of any one philosopher-king could be consistent with the goals and values of a majority of the population, or even with the goals and values of a majority of any elite stratum (the intellectuals, say, or scientists, or rich people)—except to the extent that the philosopher-king, once in power, might use propaganda or other techniques of human engineering to bring the values of the majority into line with his own. If the realities of practical politics are taken into account, it seems that anyone who might actually become a philosopher-king either would have to be a compromise candidate, a bland fellow whose chief concern would be to avoid offending anyone, or else would have to be the ruthless leader of an aggressive faction that drives its way to power. In the latter case he might be an unscrupulous person intent only on attaining power for himself (a Hitler), or he might be a sincere fanatic convinced of the righteousness of his cause (a Lenin), but either way he would stop at nothing to achieve his goals.

Thus, the citizen who might find the idea of a philosopher-king attractive should bear in mind that he himself would not select the philosopher-king, and that any philosopher-king who might come into power would probably not be the kind that he imagines or hopes for.

A further problem is that of selecting a successor when the philosopher-king dies. Each philosopher-king will have to be able to pre-select reliably a successor whose goals and values are virtually identical to his own; for, otherwise, the first philosopher-king will steer the society in one direction, the second philosopher-king will steer the society in a somewhat different direction, the third philosopher-king will steer it in yet another direction, and so forth. The result will be that the development of the society in the long term will wander at random, rather than being steered in any consistent direction or in accord with any consistent policy as to what constitute desirable or undesirable outcomes.

Historically, in absolute monarchies of any kind—the Roman Empire makes a convenient example—it has proven impossible even to ensure the succession of rulers who are reasonably competent and conscientious. Capable, conscientious rulers have alternated with those who have been irresponsible, corrupt, vicious, or incompetent. As for a long, unbroken succession of rulers, each of whom not only is competent and conscientious but also has goals and values closely approximating those of his predecessor—you can forget it. All of these arguments, by the way, apply not only to philosopher-kings but also to philosopher-oligarchs—ruling groups small enough so that Engels's "conflicts among many individual wills" do not come into play.

All the same, let's assume that it would somehow be possible to ensure the succession of a long line of philosopher-kings all of whom would govern in accord with a single, permanently stable system of values. In that event... but hold on... let's pause and take stock of the assumptions we've been making. We're assuming, among other things, that the problems of complexity, chaos, and the resistance of subordinates, also the purely technical factors that limit the options open to leaders, as well as the competitive, power-seeking groups that evolve within a society under the influence of natural selection, can all be overcome to such an extent that an all-powerful leader will be able to govern the society rationally; we're assuming that the "conflicts among many individual wills" within the society can be resolved well enough so that it will be possible to make a rational choice of leader; we're assuming that means will be found to put the chosen leader into a position of absolute power and to guarantee forever the succession of competent and conscientious leaders who will govern in accord with some stable and permanent system of values. And if the hypothetical possibility of steering a society rationally is to afford any comfort to the reader, he will have to assume that the system of values according to which the society is steered will be one that is at least marginally acceptable to himself—which is a sufficiently daring assumption.

It's now clear that we have wandered into the realm of fantasy. It is impossible to prove with mathematical certainty that the development of a society can never be guided rationally over any significant interval of time, but the series of assumptions that we've had to make in order to entertain the possibility of rational guidance is so wildly improbable that for practical purposes we can safely assume that the development of societies will forever remain beyond rational human control.[123]

VI. It's likely that the chief criticism to be leveled at this chapter will be that the writer has expended a great deal of ink and paper to prove what "everyone" already knows. Unfortunately, however, not everyone does know that the development of societies can never be subject to rational human control; and even many who would agree with that proposition as an abstract principle fail to apply the principle in concrete cases. Again and again we find seemingly intelligent people proposing elaborate schemes for solving society's problems, completely oblivious to the fact that such schemes never, never, never are carried out successfully. In a particularly fuddled excursion into fantasy written several decades ago, the noted technology critic Ivan Illich asserted that "society must be reconstructed to enlarge the contribution of autonomous individuals and primary groups to the total effectiveness of a new system of production designed to satisfy the human needs which it also determines," and that a "convivial society should be designed to allow all its members the most autonomous action by means of tools least controlled by others"[124]—as if a society could be consciously and rationally "reconstructed" or "designed." Other egregious examples of this sort of folly were provided by Arne Naess[125] and Chellis Glendinning[126] in 1989 and 1990, respectively; these are discussed in Part IV of Chapter Three of this book.

Right down to the present (2013), people who should know better have continued to ignore the fact that the development of societies can never be rationally controlled. Thus, we often find technophiles making such absurd statements as: "humanity is in charge of its own fate"; "[we will] take charge of our own evolution"; or, "people [will] seize control of the evolutionary process."[127] The technophiles want to "guide research so that technology improve[s] society"; they have created a "Singularity University" and a "Singularity Institute" that are supposed to "shape the advances and help society cope with the ramifications" of technological progress, and "make sure... that artificial intelligence... is friendly" to humans.[128]

Of course, the technophiles won't be able to "shape the advances" of technology or make sure that they "improve society" and are friendly to humans. Technological advances will be "shaped" in the long run by unpredictable and uncontrollable power-struggles among rival groups that will develop and apply technology for the sole purpose of gaining advantages over their competitors. See Chapter Two.

It's not likely that the majority of technophiles fully believe in this drivel about "shaping the advances" of technology to "improve society." In practice, Singularity University serves mainly to promote the interests of technology-oriented businessmen,[129] while the fantasies about "improving society" function as propaganda that helps to forestall public resistance to radical technological innovation. But such propaganda is effective only because many laymen are naïve enough to take the fantasies seriously.

Whatever may be the motives behind the technophiles' schemes for "improving society," other such schemes unquestionably are sincere. For recent examples, see the books by Jeremy Rifkin (2011)[130] and Bill Ivey (2012).[131] There are other examples that superficially look more sophisticated than the proposals of Rifkin and Ivey but are equally impossible to carry out in practice. In a book published in 2011, Nicholas Ashford and

Ralph P. Hall[132] "offer a unified, transdisciplinary approach for achieving sustainable development in industrialized nations. ... The authors argue for the design of multipurpose solutions to the sustainability challenge that integrate economics, employment, technology, environment, industrial development, national and international law, trade, finance, and public and worker health and safety."[133] Ashford and Hall do not intend their book to be merely an abstract speculation like Plato's *Republic*[134] or Thomas More's *Utopia*; they imagine themselves to be offering a practical program.[135]

In another example (2011), Naomi Klein proposes massive, elaborate, worldwide "planning"[136] that is supposed to bring global warming under control,[137] help with many of our other environmental problems,[138] and at the same time bring us "real democracy,"[139] "rein in"[140] the corporations, alleviate unemployment,[141] reduce wasteful consumption in rich countries[142] while allowing poor countries to continue their economic growth,[143] foster "interdependence rather than hyper-individualism, reciprocity rather than dominance and cooperation rather than hierarchy,"[144] "elegantly weav[e] all these struggles into a coherent narrative about how to protect life on earth,"[145] and overall promote a "progressive" agenda[146] so as to create a "healthy, just world."[147]

One is tempted to ask whether the schemes concocted by people like Ashford, Hall, and Klein[148] are meant as an elaborate joke of some sort; but no, the intentions of these authors are quite serious. How can they possibly believe that schemes like theirs will ever be carried out in the real world? Are they totally devoid of any practical sense about human affairs? Maybe. But Naomi Klein herself unwittingly offers a more likely explanation: "[I]t is always easier to deny reality than to watch your worldview get shattered... ."[149] The worldview of most members of the upper middle class, including most intellectuals, is deeply dependent on the existence of a thoroughly organized, culturally "advanced," large-scale society characterized by a high level of social order. It would be extremely difficult psychologically for such people to recognize that the only way to get off the road to disaster that we are now on would be through a total collapse of organized society and therefore a descent into chaos. So they cling to any scheme, however unrealistic, that promises to preserve the society on which their lives and their worldview are dependent; and one suspects that the threat to their worldview is more important to them than the threat to their lives.

NOTES

1. Redondilla, in Barja, p. 176. Free translation: "Where there is a plan for good, some evil derails it. The good comes but proves ineffective, while the evil is effective and persists."

2. Tacitus, Book III, Chapt. 18, p. 112.

3. This is somewhat of an oversimplification, but it's close enough to the truth for our purposes. See NEB (2003), Vol. 4, "Federal Reserve System," p. 712, and Vol. 8, "monetary policy," pp. 251–52; *World Book Encyclopedia,* 2011, Vol. 7, "Federal Reserve System," p. 65.

4. NEB (2003), Vol. 20, "Greek and Roman Civilizations," pp. 295–96.

5. Ibid., pp. 304–05.

6. NEB (1997), Vol. 22, "Italy," p. 195.

7. Simón Bolívar, Letter to Gen. Juan José Flores, Nov. 9, 1830, in Soriano, p. 169.

8. R. Heilbroner & A. Singer, p. 122.

9. Patterson, pp. 402–03.

10. The facts are outlined by Patterson, pp. 396–405, but the conclusions drawn from the facts are my own.

11. E.T. Brooking & P.W. Singer, p. 83.

12. Ibid. (the entire article). Manjoo (the entire book).

13. E.T. Brooking & P.W. Singer (the entire article).

14. There are at least three categories of exceptions to this rule, as noted in Kaczynski, Letter to David Skrbina: Oct. 12, 2004, Part III.A, but these exceptions have little relevance to the present chapter.

15. NEB (2003), Vol. 20, "Greek and Roman Civilizations," pp. 228–29. But see Starr, pp. 314, 315, 317, 334&n8, 350, 358.

16. NEB (2003), Vol. 20, "Germany," p. 114.

17. NEB (2003), Vol. 15, "Bismarck," p. 124. For Bismarck's career generally, see ibid., pp. 121–24; ibid., Vol. 20, "Germany," pp. 109–114; Zimmermann, Chapts. 1&7; Dorpalen, pp. 219–220, 229–231, 255–56, 259–260&n53.

18. *Constitution of the United States,* Amendments XVIII & XXI. Patterson, pp. 167–69. NEB (2003), Vol. 29, "United States of America," pp. 254–55. *Encyclopedia of American Studies* (2001), Vol. 3, "Prohibition," pp. 414–17. Okrent, p. 373. *USA Today,* Aug. 17, 2016, p. 8A. Vergano, p. 3A, says that according to Arthur Lurigio of Loyola University Chicago, "Prohibition... was unique in its widespread loathing by the populace and that opening is what enabled organized crime to gain its political footing in Chicago."

19. Naruo Uehara, p. 235.

20. Bourne, pp. 46–47.

21. E.g.: Sohail Ejaz et al., pp. 98–102 (Pakistan, medical effects);

Yukinori Okada & Susumu Wakai, pp. 236–242 (Thailand, economic and medical effects); Naruo Uehara, p. 235 (various effects, including desertification in unspecified countries); Aditya Batra (Sri Lanka, medical effects); Guillette et al., pp. 347–353 (Mexico, medical and behavioral effects); Watts (entire work) (various countries, various effects).

22. NEB (2003), Vol. 4, "Eisenhower, Dwight D(avid)," p. 405; Vol. 18, "Energy Conversion," p. 383; Vol. 29, "United Nations," p. 144.

23. A.K. Smith & C. Weiner, pp. 271, 291, 295, 310, 311, 328.

24. NEB (2003), Vol. 29, "United Nations," p. 144.

25. F. Zakaria, p. 34.

26. *The Economist*, June 18, 2011, "Move the base camp," pp. 18, 20, and "The growing appeal of zero," p. 69.

27. See Kaczynski, Letter to David Skrbina: March 17, 2005, Part I.A, point 12; and either Afterthought 3 of the Feral House edition or Appendix Five of the Fitch & Madison edition. Also, "Radioactive fuel rods: The silent threat," *The Week*, April 15, 2011, p. 13.

28. Joy, p. 239.

29. Steele, pp. 5–21. It is also claimed that a free market provides a mechanism that "automatically" maximizes the efficiency of an economy. This last contention is unproven and probably far from accurate, but the argument that excessive complexity makes rationally planned economies impossible is very strong.

30. Ibid., p. 83. Stigler, p. 113.

31. "It is 'absurd' to suppose that the information could be collected. . . ." Steele, p. 83.

32. The text of the talk can be found in Lorenz, pp. 181–84.

33. *Time*, May 5, 2008, p. 18. *The Week*, May 2, 2008, p. 35.

34. NEB (2003), Vol. 3, "chaos," p. 92.

35. Ibid., Vol. 25, "Physical Sciences, Principles of," p. 826.

36. Ibid., pp. 826–27.

37. Ibid., p. 826.

38. See Kaczynski, Letters to a German, paragraph that begins, "Among the few reliable predictions. . . ."

39. See Kelly, pp. 159ff. But Moore himself thinks the law is a "self-fulfilling prophecy," i.e., it continues to hold true only because people believe in it. Ibid., p. 162.

40. Kurzweil, e.g., pp. 351–368.

41. Russell's Paradox: Let a set be called "ordinary" if, and only if, it is not a member of itself, and let S be the set of all ordinary sets. Is S ordinary, or not?

42. See note 39.

43. Thurston, p. xviii. See also Buechler, p. 27 ("social action always produces unintended and unanticipated consequences").

44. Isaacson, p. 697.

45. Elias, p. 543n1. See also R. Heilbroner & A. Singer, p. 112 (much of economic history "follow[s] from the blind workings of the market mechanism").

46. Elias, p. 367. However, Elias continues: "But it is by no means impossible that we can make of it something more 'reasonable,' something that functions better in terms of our needs and purposes. For it is precisely in conjunction with the civilizing process that the blind dynamics of people intertwining in their deeds and aims gradually leads toward greater scope for planned intervention into both the social and individual structures—intervention based on a growing knowledge of the unplanned dynamics of these structures." But Elias does not even pretend to offer any evidence to support this statement, which is mere speculation—in contrast to his statements about the unplanned and unintended character of all earlier historical change, which are abundantly supported by his empirical studies of the ways in which European society changed in the past. What Elias suggests here looks very much like the proposal set forth at the beginning of Part III of the present chapter, and that proposal is disposed of in Part III.

When Elias claims that "we can make of [society]... something that functions better in terms of our needs and purposes," he fails to explain who this "we" is. Obviously, "we" don't all have the same purposes, and the effort to fulfill some of "our" needs (e.g., status, power) inevitably brings us into conflict with others among the "we." See Parts III and IV of this chapter.

Though the edition of Elias's book cited here is dated 2000, the content was written several decades earlier. Since that time there has been no discernible improvement in humans' capacity for "planned intervention" in the development of their societies. If anything, our statesmen seem even less in control of events than they were in the past. Elias's formative years were in the first half of the 20th century, when a belief in "progress" was still widely current. Elias seems to have been reluctant—not for rational reasons—to relinquish that belief. His remarks on that subject, ibid., pp. 462–63, are ill-advised.

47. Engels, Letter to Joseph Bloch, as referenced in our List of Works Cited. Engels of course wrote in German. The translation given here is influenced both by the English translation in *Historical Materialism* (see the List of Works Cited), pp. 294–96, and by the Spanish translation provided by Carrillo, pp. 111–12. Since Carrillo was Secretary General of the Communist Party of Spain, he presumably was learned in Engels's ideas.

48. Elias, p. 311. But see note 46, above.

49. See Kaczynski, Letter to David Skrbina: March 17, 2005, Part I.A, point 11. The problem of the commons is also called the "tragedy of the commons," and the term is often used in a narrower sense than that in which I use it here. See, e.g., Diamond, pp. 428–430. But the term is also used in the broader sense in which I apply it. E.g., *The Economist*, April 2, 2011, p. 75. Without using the

term "problem" or "tragedy of the commons," Surowiecki, p. 25, has illustrated the concept by giving several excellent examples of ways in which "individually rational decisions [can] add[ ] up to a collectively irrational result."

50. The second sentence of this paragraph is based on my notes of a conversation with Lt. Lewis, written within a couple of hours after the end of that conversation. The relevant pages are No. 04–1013 and No. 04–1016 of my Bates-numbered notes to my attorneys, which should now be in the Labadie Collection at the University of Michigan's Special Collections Library. The last sentence of the paragraph is based on my recollection (2012) of the same conversation.

51. From a speech attributed to Caesar by Sallust, *Conspiracy of Catiline,* section 51, p. 217. Roman historians commonly invented the speeches that they attributed to famous people, but the quoted statement is worth noting whether it represented Caesar's opinion or Sallust's.

52. Brathwait, quoted by Boorstin, pp. 99–100. I've taken the liberty of modernizing spelling and capitalization.

53. NEB (2003), Vol. 23, "Lincoln," p. 36.

54. Sampson, pp. 454–55. See also p. 436 (Mandela " 'was still operating under the illusion, cherished by so many revolutionaries,' complained de Klerk…, 'that possession of the levers of government enabled those in power to achieve whatever goals they wanted.' ").

55. Ibid., p. 498.

56. Rossiter, pp. 52–64.

57. Ibid., p. 54.

58. Ibid.

59. Ibid., pp. 167–68.

60. Mote, p. 98.

61. Ibid., p. 99.

62. Ibid.

63. Ibid.

64. See ibid., pp. 99–100.

65. Ibid., p. 139.

66. NEB (2003), Vol. 16, "China," p. 100. Mote, pp. 139–142.

67. NEB (2003), loc. cit.

68. Mote, p. 142. NEB (2003), loc. cit.

69. Mote, p. 142. For emperors' dates see ibid., p. 105, Chart 2. Zhezong technically became emperor in 1085, but the country was governed by a regent until approximately 1093.

70. Ibid., p. 207.

71. Ibid., p. 143.

72. Ibid., p. 207.

73. Ibid., p. 143.

74. Elias, pp. 312–344.

75. Ibid., pp. 343–44.

76. Ibid., p. 38. For an inkling of the limitations on what one very nearly absolute monarch can do today, see Goldberg, pp. 44–55 (about King Abdullah II of Jordan).

77. NEB (2003), Vol. 14, "Austria," pp. 518–520.

78. It could be argued that dissolution of the "traditional... social structures and customary restraints" was a *precondition* for the dictators' rise to power. See Selznick, p. 281n5.

79. Thurston, p. 169. For other information on the regime's inability to control its labor force, see ibid., pp. 167–172, 176, 184.

80. Ibid., p. 172. Ulam, p. 342.

81. Thurston, p. 171. There were of course other ways in which Stalin's intentions were thwarted by some combination of economic realities and ordinary people's refusal to cooperate. E.g., Stalin's program of forced collectivization wrecked Soviet agriculture, which never fully recovered. Ulam, pp. 330–37, 355–56. It might be claimed that, despite his inability to control his work-force, Stalin did exercise control to the extent that he achieved the industrialization of the Soviet Union. But industrialization had already begun under the tsars, and, given Russia's proximity to the industrialized West, its continuing industrialization was only to be expected. Thus, in industrializing Russia, Stalin merely pushed the country along its predetermined course of development; and under a capitalist regime it's likely that industrialization would have proceeded more rapidly and more efficiently.

82. Thurston, pp. 17, 57, 90, 106, 112, 147, 227–28, 233. Stalin did not plan the Terror *as it actually developed*, but Thurston's argument is insufficient to prove that Stalin did not plan to initiate a terror campaign of *some* sort, though the terror campaign that he did initiate proved uncontrollable. In other respects, and as far as *concrete facts* (as opposed to rhetoric) are concerned, Thurston's view is mostly consistent with the more traditional view of Stalin as the "mastermind of a plot to subdue the party and the nation." For all this, and for some remarks on state terrorism in general, see Appendix Five.

83. Thurston, p. 200. Ulam, pp. 445–48, 489, 521, 523.

84. Dorpalen, p. 418.

85. Rothfels, pp. 58–59.

86. Ulam, p. 447.

87. Kosthorst, pp. 108–110. Rothfels, pp. 97, 104, 227n88.

88. Hitler did purge his army to a limited extent. Thurston, p. 200. Rothfels, p. 88. But this was not comparable to the kind of thorough purge that Stalin carried out. Apparently the officers purged by Hitler were merely dismissed

from the service, not executed or imprisoned.

89. Rothfels, p. 100.

90. To mention only the three most striking examples:

(i) In 1938 some of Hitler's generals planned a coup d'état, which was to take place on the morning of September 29. The order to proceed with the coup had already been given when the announcement on September 28 of Neville Chamberlain's flight to Munich, where he was to negotiate the famous agreement, seemed to remove the rationale for the coup. Rothfels, pp. 78–79. Kosthorst, p. 10, provides the information that the order to proceed with the coup had already been given.

(ii) On March 13, 1943, Lieutenant Fabian von Schlabrendorff succeeded in planting a bomb on Hitler's plane. Rothfels, p. 99. But "the detonator cap… failed to fire. The explanation was probably that the cabin temperature in the aircraft had been sub-zero, due to a fault in the heating system, and this had affected the detonator." A. Read & D. Fisher, p. 118.

(iii) In the well-known assassination attempt of July 20, 1944, the bomb did explode but failed to disable Hitler. Often cited as the chance circumstances that saved the Führer are the fact that he happened to be leaning over a heavy oaken table at the moment when the bomb went off under the table, and the fact that someone had pushed the briefcase containing the bomb behind one of the thick wooden supports that held up the table. Far more important, however, was an error that on the part of German military officers—renowned for their technical efficiency—seems incredible: The would-be assassins neglected to provide their bomb with shrapnel. Had ample shrapnel been provided, Hitler's legs likely would have been mangled; he would have been unconscious on an operating table and unable to conduct the telephone conversation with Major Remer that quashed the coup attempt. See Gilbert, *Second World War*, pp. 557–59, and the diagram on p. 1059 of Cebrián et al.

91. Ulam, p. 474.

92. Norris, Book II, Chapter VIII, pp. 285–86. I've taken the liberty of improving the capitalization and punctuation.

93. Patterson, p. 65.

94. Dulles, pp. 73–75.

95. Ibid., p. 99.

96. Davidson, pp. 66ff.

97. R. Heilbroner & A. Singer, p. 84.

98. NEB (2003), Vol. 28, "Spain," p. 10.

99. Sueiro & Díaz Nosty, pp. 309–317. Suárez, pp. 231–33, 418, 471–72, 483–88. Payne, pp. 16–23.

100. See Payne, p. 17 ("[Franco era] un ignorante del funcionamiento de la economía—como casi todos los dictadores—y creía que se podía lidiar con ella

como lo hacía un general con su ejército: dando órdenes y directrices desde arriba sobre cómo debía comportarse.").

101. Matthews, pp. 79, 108. Horowitz, pp. 64, 127–28.

102. Matthews, pp. 76, 96–97, 337. Horowitz, pp. 46, 146–47.

103. Matthews, pp. 108, 201. Horowitz, pp. 41–84, 128, 130–32, 145, 157.

104. Matthews, pp. 83, 337–38. Horowitz, pp. 129–130, 133. Saney, pp. 19, 40n1.

105. E.g., Matthews, pp. 76, 254, 337; Horowitz, pp. 41, 46, 47, 56.

106. Horowitz, p. 120. Cf. Saney, pp. 20–21.

107. Saney, pp. 112–13.

108. This is the impression one gets from Saney, pp. 100–121. Cf. Horowitz, p. 117.

109. Saney, pp. 19–21. Horowitz, pp. 46, 48, 60, 77, 175. Steele, p. 405n17. NEB (2003), Vol. 3, "Cuba," p. 773; Vol. 29, "West Indies," pp. 735, 739.

110. Saney, pp. 19–20. Horowitz, pp. 129–134. Matthews, p. 201 ("... in so many... ways, [Castro] found that his 'utopian' ideas did not satisfy his real needs").

111. See *USA Today*, Sept. 9, 2010, p. 4A, May 10, 2011, p. 6A, and June 8–10, 2012, p. 9A; *Time*, Sept. 27, 2010, p. 11; *The Week*, April 29, 2011, p. 8; Horowitz, p. 175.

112. Horowitz, pp. 111–12, 129, 158, 161–63, 174–75.

113. See ibid., pp. 175–76.

114. Ibid., pp. 43, 77, 123.

115. Saney, p. 21.

116. Horowitz, e.g., pp. 30, 75–77, 120.

117. Other factors contributing to Cuba's economic failure were: (i) The limited natural and human resources of the island. Saney, pp. 15, 19. Horowitz, p. 145. But Singapore had negligible natural resources, yet built an impressively powerful economy. Human resources (trained technical personnel, etc.) can be created in a relatively short time, as in Japan following the Meiji Restoration. The Cubans would not have had to be as industrious or as skillful as the Singaporeans or the Japanese in order to build merely an *adequate* economy. (ii) Cuba's economic dependence on the Soviet Union. Saney, p. 21. Horowitz, pp. 77, 99, 111, 120, 128, 147. But Cuba's dependence was only a result of its failure from other causes. An economically sound nation would have been able to avoid total dependence on a single foreign power.

118. The idea of a "philosopher-king" originated with Plato (see in Buchanan: "The Republic," Book V, p. 492; Book VI), who seems to have entertained not only the notion of a single philosopher-king (ibid., Book VI, pp. 530–31), but also that of a philosopher-oligarchy (ibid., Book VII, p. 584: "... when the true philosopher kings are born in a State, one or more of them...").

From respect for the female sex, let's note that the hypothetical philosopher "king" considered in Part V of this chapter could just as well be a philosopher-queen.

119. Leuchtenburg, pp. 26, 27.

120. Ibid., p. 30.

121. Ibid., pp. 30n43, 221–22.

122. Gilbert, *European Powers*, pp. 191–92.

123. True believers in technology like Ray Kurzweil and Kevin Kelly will no doubt propose futuristic, hypertechnological solutions to the problem of rational guidance of a society. For our answer, see Appendix One.

124. Illich, pp. 10, 20.

125. Naess, pp. 92–103.

126. Glendinning, as referenced in our List of Works Cited.

127. Grossman, p. 49, col. 1, col. 3. Vance, p. 1.

128. Grossman, p. 48, col. 3. Markoff, "Ay Robot!," p. 4, col. 2, col. 3 (columns occupied entirely by advertisements are not counted).

129. See, e.g., Vance, p. 1 (Singularity University "focuses on introducing entrepreneurs to promising technologies...," etc.).

130. Rifkin, as referenced in our List of Works Cited.

131. Ivey, as referenced in our List of Works Cited.

132. Ashford & Hall, as referenced in our List of Works Cited.

133. Publisher's description located online as of March 28, 2016 at: http://yalebooks.com/book/9780300169720/technology-globalization-and-sustainable-development. The bit quoted here does truthfully describe the content of the book.

134. Plato did not regard his "Republic" as mere abstract speculation; he thought he was describing, at least to a rough approximation, a practical possibility. See in Buchanan: "The Republic," Book V, pp. 491–92; Book VI, pp. 530–31; Book VII, p. 584. But in modern times—as far as I know—Plato's "Republic" has always been treated as theoretical speculation, not as a description of a practical possibility.

135. Ashford & Hall, p. 1 ("We hope that the prescriptions discussed in this work will not be regarded as utopian.").

136. Klein, pp. 14–15.

137. Ibid., pp. 14–17.

138. Ibid., p. 15.

139. Ibid., p. 15, col. 1.

140. Ibid.; see also p. 18, col. 1 ("reining in of the market forces").

141. Ibid., pp. 15, col. 1, col. 2; 16; 21, col. 2.

142. Ibid., pp. 16; 17, col. 2.

143. Ibid., p. 16.

144. Ibid., p. 19, col. 2.

145. Ibid., p. 20, col. 1.

146. Ibid.

147. Ibid., p. 20, col. 2.

148. For a more recent (2015) example of such delusions, see Gardner, Prugh & Renner, p. 17 ("The world now needs to adopt solutions that change the entire system of production and consumption in a fundamental manner.... This... requires large-scale social, economic, and political engineering...").

149. Klein, p. 18, col. 1.

CHAPTER TWO

Why the Technological System Will Destroy Itself

> We were recently entertained by a naïve fable of the happy arrival of the 'end of history,' of the overflowing triumph of an all-democratic bliss; the ultimate global arrangement had supposedly been attained. But we all see and sense that something very different is coming, something new, and perhaps quite stern.
>
> — Aleksandr Solzhenitsyn[1]

> Power is in nature the essential measure of right.
>
> — Ralph Waldo Emerson[2]

I. Most of the arguments set forth elsewhere in this book are reasonably solid, but in the present chapter we go out on a limb both in making assumptions and in drawing inferences from them. We think our assumptions and inferences contain at least as much truth as they need to contain for the purpose of reaching certain probable conclusions about the future of human society, but we acknowledge that rational disagreement with our reasoning is possible. Two things, however, can be definitely asserted: first, that our assumptions and inferences are reasonably accurate as applied to the development up to the present time of large-scale, complex societies; second, that anyone who wants to understand the likely future development of modern society will have to give careful attention to problems of the kind that are raised by the arguments of this chapter.

Though we focus here on the processes of competition and natural selection[3] as they operate in complex societies, it is important to avoid confusing our viewpoint with the (now largely defunct) philosophy known as "Social Darwinism." Social Darwinism didn't merely call attention to natural selection as a factor in the development of societies; it also assumed that the winners in the contest of "survival of the fittest" were better, more desirable human beings than the losers were:

> [T]he competitive struggle of business was viewed as a contest in which the survivors were the 'fittest'—not merely as businessmen, but as champions of civilization itself. Hence businessmen transformed their sense of material superiority into a sense of moral and intellectual superiority. ... Social Darwinism became a means of excusing as well as explaining the competitive process from which some emerged with power and some were ground into poverty.[4]

Here our purpose is merely to describe the role that natural selection plays in the development of societies. We do not mean to suggest any favorable value-judgment concerning the winners in the struggle for power.

II. This chapter deals with self-propagating systems. By a self-propagating system (self-prop system for short) we mean a system that tends to promote its own survival and propagation. A system may propagate itself in either or both of two ways: The system may indefinitely increase its own size and/or power, or it may give rise to new systems that possess some of its own attributes.

The most obvious examples of self-propagating systems are biological organisms. *Groups* of biological organisms can also constitute self-prop systems; e.g., wolf packs or hives of honeybees. Particularly important for our purposes are self-prop systems that consist of groups of human beings. For example, nations, corporations, labor unions, churches, and political parties; also some groups that are not clearly delimited and lack formal organization, such as schools of thought, social networks, and subcultures. Just as wolf-packs and beehives are self-propagating without any conscious intention on the part of wolves or bees to propagate their packs or their hives, there is no reason why a human group cannot be self-propagating independently of any intention on the part of the individuals who comprise the group.

If A and B are systems of any kind (self-propagating or not), and if A is a functioning component of B, then we will call A a *subsystem* of B, and we will call B a *supersystem* of A. For example, in human hunting-and-gathering societies, nuclear families[5] belong to bands, and bands often are organized into tribes. Nuclear families, bands, and tribes are all self-prop systems. The nuclear family is a subsystem of the band, the band is a subsystem of the tribe, the tribe is a supersystem of each band that belongs to it, and each band is a supersystem of every nuclear family that belongs to that band. It is also true that each nuclear family is a subsystem of the tribe and that the tribe is a supersystem of every nuclear family that belongs to a band that belongs to the tribe.

The principle of natural selection is operative not only in biology, but in any environment in which self-propagating systems are present. The principle can be stated roughly as follows:

Those self-propagating systems having the traits that best suit them to survive and propagate themselves tend to survive and propagate themselves better than other self-propagating systems.

This of course is an obvious tautology, so it tells us nothing new. But it can serve to call our attention to factors that we might otherwise overlook.

We are about to advance several propositions that are not tautologies. We can't prove these propositions, but they are intuitively plausible and they seem consistent with the observable behavior of self-propagating systems as represented by biological organisms and human (formal or informal) organizations. In short, we believe these propositions to be true, or as close to the truth as they need to be for present purposes.

Proposition 1. In any environment that is sufficiently rich, self-propagating systems will arise, and natural selection will lead to the evolution of self-propagating systems having increasingly complex, subtle, and sophisticated means of surviving and propagating themselves.

It needs to be emphasized that natural selection doesn't merely act in simple ways, as by making the legs of deer longer so that they can run faster or giving arctic mammals thicker coats of fur so that they can stay warm. Natural selection can also lead to the development of complex structures such as the human eye or heart, and to systems of far greater complexity that still are not fully understood, such as the human immune system or nervous system. We maintain that natural selection can lead to equally complex and subtle developments in self-prop systems consisting of human groups.

Natural selection operates relative to particular periods of time. Let's start at some given point in time that we can call Time Zero. Those self-prop systems that are most likely to survive (or have surviving progeny) at five years from Time Zero are those that are best suited to survive and propagate themselves (in competition[6] with other self-prop systems) during the five-year period following Time Zero. These will not necessarily be the same as those self-prop systems that, in the absence of competition during the five-year period, would be best suited to survive and propagate themselves during the thirty years following Time Zero. Similarly, those systems best suited to survive competition during the first thirty years following Time Zero are not necessarily those that, in the absence of competition during the thirty-year period, would be best suited to survive and propagate themselves for two hundred years. And so forth.

For example, suppose a forested region is occupied by a number of small, rival kingdoms. Those kingdoms that clear the most land for agricultural use can plant more crops and therefore can support a larger population than other kingdoms. This gives them a military advantage over their rivals. If any kingdom restrains itself from excessive forest-clearance out of concern for the long-term consequences, then that kingdom places itself at a military disadvantage and is eliminated by the more powerful kingdoms. Thus the region comes to be dominated by kingdoms that cut down their forests recklessly. The resulting deforestation leads eventually to ecological disaster and therefore to the collapse of all the kingdoms. Here a trait that is advantageous or even indispensable for a kingdom's short-term survival—recklessness in cutting trees—leads in the long term to the demise of the same kingdom.[7]

This example illustrates the fact that, where a self-prop system exercises foresight,[8] in the sense that concern for its own long-term survival and propagation leads it to place limitations on its efforts for short-term survival and propagation, the system puts itself at a competitive disadvantage relative to those self-prop systems that pursue short-term survival and propagation without restraint. This leads us to

Proposition 2. In the short term, natural selection favors self-propagating systems that pursue their own short-term advantage with little or no regard for long-term consequences.

A corollary to Proposition 2 is

Proposition 3. Self-propagating subsystems of a given supersystem tend to become dependent on the supersystem and on the specific conditions that prevail within the supersystem.

This means that between the supersystem and its self-prop subsystems, there tends to develop a relationship of such a nature that, in the event of the destruction of the supersystem or of any drastic acceleration of changes in the conditions prevailing within the supersystem, the subsystems can neither survive nor propagate themselves.

A self-prop system with sufficient foresight would make provision for its own or its descendants' survival in the event of the collapse or destabilization of the supersystem. But as long as the supersystem exists and remains more or less stable, natural selection favors those subsystems that take fullest advantage of the opportunities available within the supersystem, and disfavors those subsystems that "waste" some of their resources in preparing themselves to survive the eventual destabilization of the supersystem. Under these conditions, self-prop systems will tend very strongly to become incapable of surviving the destabilization of any supersystem to which they belong.

Like the other propositions put forward in this chapter, Proposition 3 has to be applied with a dose of common sense. If the supersystem in question is weak and loosely organized, or if it has no more than a modest effect on the conditions in which its subsystems exist, the subsystems may not become strongly dependent on the supersystem. Among hunter-gatherers in some (not all) environments, a nuclear family would be able to survive and propagate itself independently of the band to which it belongs. Because tribes of hunter-gatherers are loosely organized it seems certain that in most cases a hunting-and-gathering band would be able to survive independently of the tribe to which it belongs. Many labor unions might be able to survive the demise of a confederation of labor unions such as the AFL-CIO, because such an event might not fundamentally affect the conditions under which labor unions have to function. But labor unions could not survive the demise of modern industrial society, or even the demise merely of the legal and constitutional framework that makes it possible for labor unions as we know them to operate. Nor would many present-day business enterprises survive without modern industrial society. Domestic sheep, if deprived of human protection, would soon be killed off by predators. And so forth.

Clearly a system cannot be effectively organized for its own survival and propagation unless the different parts of the system can promptly communicate with one another and lend aid to one another. In order to operate effectively throughout a given geographical region, a self-prop system must be able to receive prompt information from, and take prompt action within, every part of the region.[9] Consequently,

Proposition 4. Problems of transportation and communication impose a limit on the size of the geographical region over which a self-prop system can extend its operations.

Human experience suggests:

Proposition 5. The most important and the only consistent limit on the size of the geographical regions over which self-propagating human groups extend their operations is the limit imposed by the available means of transportation and communication. In other words, while not all self-propagating human groups tend to extend their operations over a region of maximum size, natural selection tends to produce *some* self-propagating human groups that operate over regions approaching the maximum size allowed by the available means of transportation and communication.

Propositions 4 and 5 can be seen operating in human history. Primitive bands or tribes usually have territories that they "own," but these are relatively small because human feet are the only means of transportation available to these societies. However, primitives who have numerous horses and live in open country over which horses can travel freely, like the Plains Indians of North America, can hold much larger territories. Pre-industrial civilizations built empires that extended over vast distances, but these empires actively created, if they did not already have, relatively rapid means of transportation and communication.[10] Such empires grew to a certain geographical size, after which they stopped growing and, in many cases, became unstable; that is, they tended to break up into smaller political units. It is probable that these empires stopped growing and became unstable because they were at the limit of what was possible with the existing means of transportation and communication.[11]

Today there is quick transportation and almost instant communication between any two parts of the world. Hence,

Proposition 6. In modern times, natural selection tends to produce some self-propagating human groups whose operations span the entire globe. Moreover, even if human beings are some day replaced by machines or other entities, natural selection will still tend to produce some self-propagating systems whose operations span the entire globe.

Current experience strongly confirms this proposition: We see global "superpowers," global corporations, global political movements, global religions, global criminal networks. Proposition 6, we argue, is not dependent on any particular traits of human beings but only on the general properties of self-prop systems, so there is no reason to doubt that the proposition will remain true if and when humans are replaced by other entities: As long as rapid, worldwide transportation and communication remain available, natural selection will tend to produce or maintain self-prop systems whose operations span the entire globe.

Let's refer to such systems as *global* self-prop systems. Instant worldwide communications are still a relatively new phenomenon and their full consequences have yet to be developed; in the future we can expect global self-prop systems to play an even more important role than they do today.

Proposition 7. Where (as today) problems of transportation and communication do not constitute effective limitations on the size of the geographical regions over which self-propagating systems operate, natural selection tends to create a world in which power is mostly concentrated in the possession of a relatively small number of global self-propagating systems.

This proposition too is suggested by human experience. But it's easy to see why the proposition should be true independently of anything specifically human: Among global self-prop systems, natural selection will favor those that have the greatest power; global or other large-scale self-prop systems that are weaker will tend to be eliminated or subjugated. Small-scale self-prop systems that are too numerous or too subtle to be noticed individually by the dominant global self-prop systems may retain more or less autonomy, but each of them will have influence only within some very limited sphere. It may be answered that a coalition of small-scale self-prop systems could challenge the global self-prop systems, but if small-scale self-prop systems organize themselves into a coalition having worldwide influence, then the coalition will itself be a global self-prop system.

We can speak of the "world-system," meaning all things that exist on Earth, together with the functional relations among them. The world-system probably should not be regarded as a self-prop system, but whether it is or not is irrelevant for present purposes.

To summarize, then, the world-system is approaching a condition in which it will be dominated by a relatively small number of extremely powerful global self-prop systems. These global systems will compete for power—as they must do in order to have any chance of survival—and they will compete for power *in the short term*, with little or no regard for long-term consequences (Proposition 2). Under these conditions, intuition tells us that desperate competition among the global self-prop systems will tear the world-system apart.

Let's try to formulate this intuition more clearly. For some hundreds of millions of years the terrestrial environment has had some degree of stability, in the sense that conditions on Earth, though variable, have remained within limits that have allowed the evolution of complex life-forms such as fishes, amphibians, reptiles, birds, and mammals. In the immediate future, all self-prop systems on this planet, including self-propagating human groups and any purely machine-based systems derived from them, will have evolved while conditions have remained within these limits, or at most within somewhat wider ones. By Proposition 3, the Earth's self-prop systems will have become dependent for their survival on the fact that conditions have remained within these limits. Large-scale self-prop human groups, as well as any purely machine-based self-prop systems, will be dependent also on conditions of more recent origin relating to the way the world-system is organized; for example, conditions relating to economic relationships. The rapidity with which these conditions change must remain within certain limits, else the self-prop systems will not survive.

This doesn't mean that all of the world's self-prop systems will die if future conditions, or the rapidity with which they change, slightly exceed some of these limits, but it does mean that if conditions go far enough beyond the limits many self-prop systems are likely to die, and if conditions ever vary wildly enough outside the limits, then, with near certainty, all of the world's more complex self-prop systems will die without progeny.

With several self-prop systems of global reach, armed with the colossal might of modern technology and competing for immediate power while exercising no restraint from concern for long-term consequences, it is extremely difficult to imagine that conditions on this planet will not be pushed far outside all earlier limits and batted around so erratically that for any of the Earth's more complex self-prop systems, including complex biological organisms, the chances of survival will approach zero.

Notice that the crucial new factor here is the availability of rapid, worldwide transportation and communication, as a consequence of which there exist global self-prop systems. There is another way of seeing that this situation will lead to radical disruption of the world-system. Students of industrial accidents know that a system is most likely to suffer a catastrophic breakdown when (i) the system is highly complex (meaning that small disruptions can produce unpredictable consequences), and (ii) tightly coupled (meaning that a breakdown in one part of the system spreads quickly to other parts).[12] The world-system has been highly complex for a long time. What is new is that the world-system is now tightly coupled. This is a result of the availability of rapid, worldwide transportation and communication, which makes it possible for a breakdown in any one part of the world-system to spread to all other parts. As technology progresses and globalization grows more pervasive, the world-system becomes ever more complex and more tightly coupled, so that a catastrophic breakdown has to be expected sooner or later.

It will perhaps be argued that destructive competition among global self-prop systems is not inevitable: A single global self-prop system might succeed in eliminating all of its competitors and thereafter dominate the world alone; or, because global self-prop systems would be relatively few in number, they could come to an agreement among themselves whereby they would refrain from all dangerous or destructive competition. However, while it is easy to talk about such an agreement, it is vastly more difficult actually to conclude one and enforce it. Just look: The world's leading powers today have not been able to agree on the elimination of war or of nuclear weapons, or on the limitation of emissions of carbon dioxide.

But let's be optimistic and assume that the world has come under the domination of a single, unified system, which may consist of a single global self-prop system victorious over all its rivals, or may be a composite of several global self-prop systems that have bound themselves together through an agreement that eliminates all destructive competition among them. The resulting "world peace" will be unstable for three separate reasons.

First, the world-system will still be highly complex and tightly coupled. Students of these matters recommend designing into industrial systems such safety features as "decoupling," that is, the introduction of "barriers" that prevent malfunctions in one part of a system from spreading to other parts.[13] Such measures may be feasible, at least in theory, in any relatively limited subsystem of the world-system, such as a chemical factory, a nuclear power-plant, or a banking system, though Perrow is not optimistic that even these limited systems will ever be consistently redesigned throughout our society to minimize the risk of breakdowns within the individual systems.[14] In regard to the world-system as a whole, we noted above that it grows ever more complex and more tightly coupled. To reverse this process and "decouple" the world-system would require the design, implementation, and enforcement of an elaborate plan that would regulate in detail the political and economic development of the entire world. For reasons explained at length in Chapter One of this book, no such plan will ever be carried out successfully.

Second, prior to the arrival of "world peace" and for the sake of their own survival and propagation, the self-prop subsystems of a given global self-prop system (their supersystem) will have put aside, or at least moderated, their mutual conflicts in order to present a united front against any immediate external threats or challenges to the supersystem (which are also threats or challenges to themselves). In fact, the supersystem would never have been successful enough to become a global self-prop system if competition among its most powerful self-prop subsystems had not been moderated.

But once a global self-prop system has eliminated its competitors, or has entered into an agreement that frees it from dangerous competition from other global self-prop systems, there will no longer be any *immediate* external threat to induce unity or a moderation of conflict among the self-prop subsystems of the global self-prop system. In view of Proposition 2—which tells us that self-prop systems will compete with little regard for long-term consequences—unrestrained and therefore destructive competition will break out among the most powerful self-prop subsystems of the global self-prop system in question.[15]

Benjamin Franklin pointed out that "the great affairs of the world, the wars, revolutions, etc. are carried on and effected by parties." Each of the "parties," according to Franklin, is pursuing its own collective advantage, but "as soon as a party has gained its general point"—and therefore, presumably, no longer faces immediate conflict with an external adversary—"each member becomes intent upon his particular interest, which, thwarting others, breaks that party into divisions and occasions... confusion."[16]

History does generally confirm that when large human groups are not held together by any immediate external challenge, they tend strongly to break up into factions that compete against one another with little regard for long-term consequences.[17] What we are arguing here is that this does not apply only to human groups, but expresses a tendency of self-propagating systems in general as they develop under the influence of natural selection. Thus, the tendency is independent of any flaws of character peculiar to human beings, and the tendency will persist even if humans are "cured" of their purported defects or (as many technophiles envision) are replaced by intelligent machines.

Third, let's nevertheless assume that the most powerful self-prop subsystems of global self-prop systems will not begin to compete destructively when the external challenges to their supersystems have been removed. There yet remains another reason why the "world peace" that we've postulated will be unstable.

By Proposition 1, within the "peaceful" world-system new self-prop systems will arise that, under the influence of natural selection, will evolve increasingly subtle and sophisticated ways of evading recognition—or, once they are recognized, evading suppression—by the dominant global self-prop systems. By the same process that led to the evolution of global self-prop systems in the first place, new self-prop systems of greater and greater power will develop until some are powerful enough to challenge the existing global self-prop systems, whereupon destructive competition on a global scale will resume.

For the sake of clarity we have described the process in simplified form, as if a world-system relatively free of dangerous competition would *first* be established and afterward would be undone by new self-prop systems that would arise. But it's more likely that new self-prop systems will be arising all along to challenge the existing global self-prop systems, and will prevent the hypothesized "world peace" from ever being consolidated in the first place. In fact, we can see this happening before our eyes.[18] The most crudely obvious of the (relatively) new self-prop systems are those that challenge law and order head on, such as terrorist networks and hackers' groups,[19] as well as frankly criminal enterprises[20] that make no pretense of idealistic motives. Drug cartels have disrupted the normal course of political life in Mexico;[21] terrorists did the same in the United States with the attack of September 11, 2001, and they are continuing to do so, much more drastically, in countries like Iraq. Self-prop systems of the purely lawless type even have the potential to take control of important nations, as drug cartels arguably have come close to doing in Kenya.[22] Political "machines" are not necessarily to be classified as criminal enterprises, but they ordinarily are more or less corrupt and tainted with illegal activity,[23] and they do challenge, or even take over, the "legitimate" structure of government.

Probably more significant for the present and the near future are emerging self-prop systems that use entirely legal methods, or at least keep their use of illegal methods to the minimum necessary for their purposes, and justify those methods with a claim, not totally outrageous, that their actions are necessary for the fulfillment of some widely accepted ideal such as "democracy," "social justice," "prosperity," "morality," or religious principles. In Israel, the ultra-orthodox sect—strictly legal—has become surprisingly powerful and seriously threatens to subvert the values and objectives of the hitherto secular state.[24] The great corporations, as we know them today, are a relatively recent (and perfectly legal) development; in the U.S. they date only from the latter half of the 19th century.[25] New corporations are continually being formed, and some grow powerful enough to challenge the older enterprises. During the last several decades many corporations have become international, and their power has begun to rival that of nation-states.[26]

A subordinate system that a government creates for its own purposes can turn into a self-prop system in its own right, and may even become dominant over the government. Thus, bureaucracies commonly are concerned more with their own power and security than with the fulfillment of their public responsibilities. "[E]very... bureaucracy develops a tendency to preserve itself, to fatten itself parasitically. It also develops a tendency to become a power in and of itself, autonomous, over which governments lose all real control."[27] In the Soviet Union, the bureaucracy became the dominant power.[28] A nation's military establishment often acquires a considerable degree of autonomy and then supplants the government as the dominant political force in the country. Nowadays the undisguised military coup seems less popular than it once was, and politically sophisticated generals prefer to exercise their power behind the scenes while allowing a facade of civilian government to function. When the generals find it necessary to intervene overtly they claim to be acting in favor of "democracy" or some such ideal. This type of military dominance can be seen today in Pakistan and Egypt.[29]

Two competing, entirely legal self-prop systems that have arisen in the U.S. during the last few decades are the politically correct left and the dogmatic right (not to be confused with the liberals and conservatives of earlier times in America). This book is not the place to speculate about the outcome of the struggle between these two forces; suffice it to say that in the long run their bitter conflict may do more to prevent the establishment of a lastingly peaceful world order than all the bombs of Al Qaeda and all the murders of the Mexican drug gangs.

People who find it difficult to face harsh realities will hope for a way to design and construct a world-system in which the processes that lead to destructive competition will not occur. But in Chapter One we've explained why no such project can ever be successfully carried out in practice. It may be objected that a mammal (or other complex biological organism) is a self-prop system that is a composite of millions of other self-prop systems, namely, the cells of its own body. Yet (unless and until the animal gets cancer) no destructive competition arises among cells or groups of cells within the animal's body. Instead, all the cells loyally serve the interests of the animal as a whole. Moreover, no external threat to the animal is necessary to keep the cells faithful to their duty. There is (it may be argued) no reason why the world-system could not be as well organized as the body of a mammal, so that no destructive competition would arise among its self-prop subsystems.

But the body of a mammal is a product of hundreds of millions of years of evolution through natural selection. This means that it has been created through a process of trial and error involving many millions of successive trials. If we suppose the duration of a generation to be a period of time Δ, those members of the first generation that contributed to the second generation by producing offspring were only those that passed the test of selection over time Δ. Those lineages[30] that survived to the third generation were only those that passed the test of selection over time 2Δ. Those lineages that survived to the fourth generation were only those that passed the test of selection over time 3Δ. And so forth. Those lineages that survived to the Nth generation were only those that passed the test of selection over the time-interval $(N-1)\Delta$ as well as the test of selection over every shorter time-interval. Though the foregoing explanation is grossly simplified, it shows that in order to have survived up to the present, a lineage of organisms has to have passed the test of selection many millions of times and over all time-intervals, short, medium, and long. To put it another way, the lineage has had to pass through a series of many millions of filters, each of which has allowed the passage only of those lineages that were "fittest" (in the Darwinian sense) to survive over time-intervals of widely varying length. It is only through this process that the body of a mammal has evolved, with its incredibly subtle and complex mechanisms that promote the survival of the animal's lineage at short, medium, and long term. These mechanisms include those that prevent destructive competition among cells or groups of cells within the animal's body.

Also highly important is the large number of individuals in each generation of a biological organism. A species that has had a close brush with extinction may at some point have been reduced to a few thousand individuals, but any mammalian species, through almost all of its evolutionary history since its first appearance as a multi-celled organism, has had millions of individuals in each generation from among which the "fittest" have been selected.

But once self-propagating systems have attained global scale, two crucial differences emerge. The first difference is in the number of individuals from among which the "fittest" are selected. Self-prop systems sufficiently big and powerful to be plausible contenders for global dominance will probably number in the dozens, or possibly in the hundreds; they certainly will not number in the millions. With so few individuals from among which to select the "fittest," it seems safe to say that the process of natural selection will be inefficient in promoting the fitness for survival of the dominant global self-prop systems.[31] It should also be noted that among biological organisms, species that consist of a relatively small number of large individuals are more vulnerable to extinction than species that consist of a large number of small individuals.[32] Though the analogy between biological organisms and self-propagating systems of human beings is far from perfect, still the prospect for viability of a world-system based on the dominance of a few global self-prop systems does not look encouraging.

The second difference is that in the absence of rapid, worldwide transportation and communication, the breakdown or the destructive action of a small-scale self-prop system has only local repercussions. Outside the limited zone where such a self-prop system has been active there will be other self-prop systems among which the process of evolution through natural selection will continue. But where rapid, worldwide transportation and communication have led to the emergence of global self-prop systems, the breakdown or the destructive action of any one such system can shake the whole world-system. Consequently, in the process of trial and error that is evolution through natural selection, it is highly probable that after only a relatively small number of "trials" resulting in "errors," the world-system will break down or will be so severely disrupted that none of the world's larger or more complex self-prop systems will be able to survive. Thus, for such self-prop systems, the trial-and-error process comes to an end; evolution through natural selection cannot continue long enough to create global self-prop systems possessing the subtle and sophisticated mechanisms that prevent destructive internal competition within complex biological organisms.

Meanwhile, fierce competition among global self-prop systems will have led to such drastic and rapid alterations in the Earth's climate, the composition of its atmosphere, the chemistry of its oceans, and so forth, that the effect on the biosphere will be *devastating*. In Part IV of the present chapter we will carry this line of inquiry further: We will argue that if the development of the technological world-system is allowed to proceed to its logical conclusion, then in all probability the Earth will be left a dead planet—a planet on which nothing will remain alive except, maybe, some of the simplest organisms—certain bacteria, algae, etc.—that are capable of surviving under extreme conditions.

* * *

The theory we've outlined here provides a plausible explanation for the so-called Fermi Paradox. It is believed that there should be numerous planets on which technologically advanced civilizations have evolved, and which are not so remote from us that we could not by this time have detected their radio transmissions. The Fermi Paradox consists in the fact that our astronomers have never yet been able to detect any radio signals that seem to have originated from an intelligent extraterrestrial source.[33]

According to Ray Kurzweil, one common explanation of the Fermi Paradox is "that a civilization may obliterate itself once it reaches radio capability." Kurzweil continues: "This explanation might be acceptable if we were talking about only a few such civilizations, but [if such civilizations have been numerous], it is not credible to believe that every one of them destroyed itself."[34] Kurzweil would be right if the self-destruction of a civilization were merely a matter of chance. But there is nothing implausible about the foregoing explanation of the Fermi Paradox if there is a process common to all technologically advanced civilizations that consistently leads them to self-destruction. Here we've been arguing that there *is* such a process.

III. Our discussion of self-propagating systems merely describes in general and abstract terms what we see going on all around us in concrete form: Organizations, movements, ideologies are locked in an unremitting struggle for power. Those that fail to compete successfully are eliminated or subjugated.[35] The struggle is almost exclusively for power in the short term;[36] the competitors show scant concern even for their own long-term survival,[37] let alone for the welfare of the human race or of the biosphere. That's why nuclear weapons have not been banned, emissions of carbon dioxide have not been reduced to a safe level, the Earth's resources are being exploited at an utterly reckless rate, and no limitation has been placed on the development of powerful but dangerous technologies.

The purpose of describing the process in general and abstract terms, as we've done here, is to show that what is happening to our world is not accidental; it is not the result of some chance conjunction of historical circumstances or of some flaw of character peculiar to human beings. Given the nature of self-propagating systems in general, the destructive process that we see today is made inevitable by a combination of two factors: the colossal power of modern technology and the availability of rapid transportation and communication between any two parts of the world.

Recognition of this may help us to avoid wasting time on naïve efforts to solve our current problems. For example, on efforts to teach people to conserve energy and resources. Such efforts accomplish nothing whatever.

It seems amazing that those who advocate energy conservation haven't noticed what happens: As soon as some energy is freed up by conservation, the technological world-system gobbles it up and demands more. No matter how much energy is provided, the system always expands rapidly until it is using all available energy, and then it demands still more. The same is true of other resources. The technological world-system infallibly expands until it reaches a limit imposed by an insufficiency of resources, and then it tries to push beyond that limit regardless of consequences.

This is explained by the theory of self-propagating systems: Those organizations (or other self-prop systems) that least allow respect for the environment to interfere with their pursuit of power here and now, tend to acquire more power than those that limit their pursuit of power from concern about what will happen to our environment fifty years from now, or even ten years. (Proposition 2.) Thus, through a process of natural selection, the world comes to be dominated by organizations that make maximum possible use of all available resources to augment their own power without regard to long-term consequences.

Environmental do-gooders may answer that if the public has been persuaded to take environmental concerns seriously it will be disadvantageous in terms of natural selection for an organization to abuse the environment, because citizens can offer resistance to environmentally reckless organizations. For example, people might refuse to buy products manufactured by companies that are environmentally destructive. However, human behavior and human attitudes can be manipulated. Environmental damage can be shielded, up to a point, from public scrutiny; with the help of public-relations firms, a corporation can persuade people that it is environmentally responsible; advertising and marketing techniques can give people such an itch to possess a corporation's products that few individuals will refuse to buy them from concern for the environment; computer games, electronic social networking, and other mechanisms of escape keep people absorbed in hedonistic pursuits so that they don't have time for environmental worries. More importantly, people are made to see themselves as utterly dependent on the products and services provided by the corporations. Because people have to earn money to buy the products and services on which they are dependent, they need jobs. Economic growth is necessary for the creation of jobs, therefore people accept environmental damage when it is portrayed as a price that must be paid for economic growth. Nationalism too is brought into play both by corporations and by governments. Citizens are made to feel that outside forces are threatening: "The Chinese will get ahead of us if we don't increase our rate of economic growth. Al Qaeda will blow us up if we don't improve our technology and our weaponry fast enough."

These are some of the tools that organizations use to counter environmentalists' efforts to arouse public concern; similar tools can help to blunt other forms of resistance to the organizations' pursuit of power. The organizations that are most successful in blunting public resistance to their pursuit of power tend to increase their power more rapidly than organizations that are less successful in blunting public resistance. Thus, through a process of natural selection, there evolve organizations that possess more and more sophisticated and effective means of blunting public resistance to their power-seeking activities, whatever the degree of environmental damage involved. Because such organizations have great wealth at their disposal, environmentalists do not have the resources to compete with them in the propaganda war.[38]

This is the reason, or an important part of the reason,[39] why attempts to teach people to be environmentally responsible have done so little to slow the destruction of our environment. And again—note well—the process we've described is not contingent on any accidental set of circumstances or on any defect in human character. Given the availability of advanced technology, the process inevitably accompanies the action of natural selection upon self-propagating systems.

IV. People who know something about the biological past of the Earth and see what the technological system is doing to our planet speak of a "sixth mass extinction," which they think is now in progress. Apparently they envision something like the extinction event at the end of the Cretaceous period, when the dinosaurs died out: They assume that many kinds of complex organisms will survive, and the species that become extinct will be replaced by complex organisms of a different kind, just as the dinosaurs were replaced by mammals.[40] Here we argue that this (relatively) comforting assumption is unjustified, because the extinction event that has now begun is of a fundamentally different kind than all of the previous mass extinctions that have occurred on this planet.

So far as is known, each previous mass extinction has resulted from the arrival of some one major disruptive factor, or at most perhaps two or three such factors.[41] Thus, it is widely believed that the dinosaurs were wiped out by the impact of an asteroid that kicked up colossal clouds of dust. These obstructed the light of the Sun, cooling the planet and interfering with photosynthesis.[42] Presumably, mammals were better able to survive under these conditions than the dinosaurs were. There are paleontologists who argue that some species of dinosaurs survived for as long as a million years after the impact of the asteroid, hence, that the asteroid alone was not enough to account for all of the extinctions that occurred at the end of the Cretaceous. The dinosaurs, they maintain, must have been finished off by some other factor—perhaps a prolonged period of unusual volcanic activity that continued to darken the atmosphere.[43] In any case, no one claims that more than a very few such factors—all of them simple, blind forces—were involved in the extinction of the dinosaurs or in other, previous mass extinctions.

In contrast to these earlier events, the extinction event that is now under way is not the work of a single blind force or even of two or three or ten such forces. Instead, it is the work of a multiplicity of intelligent, living forces. These are human organizations, self-prop systems that assiduously pursue their own short-term advantage without scruple and without concern for long-term consequences. In doing so they leave no stone unturned, no possibility untested, no avenue unexplored in their unremitting drive for power.

This can be compared to what happens in biology: In the course of evolution organisms develop means of exploiting every opportunity, utilizing every resource, and invading every corner where life is possible at all. Scientists have been surprised to discover living organisms surviving, and in some cases even thriving, in locations where there seemingly is nothing on which they could support themselves. There are communities of bacteria, worms, molluscs, and crustaceans that flourish near hydrothermal vents so deep in the ocean that no sunlight whatever can reach them and the downward drift of nutrients from the surface is entirely inadequate. Some of these creatures actually use hydrogen sulfide—to most organisms a deadly poison—as a source of energy.[44] Elsewhere there are bacteria that live a hundred feet beneath the seafloor in an environment almost completely devoid of nutrients.[45] Other bacteria nourish themselves on nothing more than "bare rock and water" at depths of up to 1.7 miles beneath the surface of the continents.[46] Everyone knows that there are organisms called parasites that find a home within other organisms, but many people may be surprised to learn that there are parasites that live in or on other parasites; in fact, there are parasites of parasites of parasites of parasites.[47]

> So, naturalists observe, a flea
> Has smaller fleas that on him prey;
> And these have smaller still to bite 'em,
> And so proceed *ad infinitum*.[48]

Needless to say, there do exist limits to the conditions under which life can survive. E.g., it has been questioned whether there can ever be a "general mechanism by which any conventional protein could be made stable and functional at temperatures above 100° C."[49] Yet some organisms do live at temperatures as high as 113° C., though none is known to survive and reproduce at a higher temperature.[50]

Like biological organisms, the world's leading human self-prop systems exploit every opportunity, utilize every resource, and invade every corner where they can find anything that will be of use to them in their endless search for power. And as technology advances, more and more of what formerly seemed useless turns out to be useful after all, so that more and more resources are extracted, more and more corners are invaded, and more and more destructive consequences follow. For example:

When humans made no use of metals other than iron meteorites, or nuggets of gold or copper that might be found by chance, the only mining activity consisted in the digging-out of rocks such as flint or obsidian that were used to make tools. But once people had learned to utilize metals on a large scale the destructive effects of mining became evident. Certainly by the 16th century, and probably much earlier, it was clearly recognized that mining poisoned streams and rivers and ruined the countryside where it occurred.[51] But in those days mining affected only a few districts where there were known deposits of relatively high-grade ore, and people who lived elsewhere probably never gave a thought to the damage caused by the extraction of metals. In recent times, however, more sophisticated means of detecting deposits of valuable minerals have been devised,[52] as well as methods for utilizing low-grade ores that formerly were left undisturbed because the extraction of metal from them was too difficult to be profitable.[53] As a result of these developments mining activities have continually invaded new areas, and severe environmental damage has followed.[54] It is said that the water flowing out of many old mining sites is so heavily contaminated that it will have to be treated "forever" to remove the toxic metals.[55] Of course, it won't be treated forever, and when the treatment stops, rivers will be irremediably poisoned.

Mining activities are invading still other areas because new uses have been found for elements that several decades ago had few if any practical applications. Most of the "rare earth" elements were of limited utility before the middle of the 20th century, but they are now considered indispensable for many purposes.[56] The rare earth neodymium, for example, is needed in large quantities for the lightweight permanent magnets used in wind turbines.[57] Unfortunately, most deposits of rare earths contain radioactive elements, hence the mining of these metals generates radioactive waste.[58]

The mining of rare earths also leads to other environmental problems, similar to those that are characteristic of mining generally.[59]

In quantitative terms, at least, uranium was of little importance prior to the development of atomic weapons and nuclear power-plants; it is now mined on a large scale. Relatively small amounts of arsenic were no doubt sufficient for medical applications and for the manufacture of rat poison and artists' pigments, but today the element is used in large quantities, e.g., to harden lead alloys and as a wood preservative. Fence posts treated with cupric arsenate are extremely common in the western United States[60]—there must be many millions of them. These posts last far longer than untreated ones, but they are not indestructible. They will eventually disintegrate, and when they do the arsenic they contain will spread through our environment. Large-scale mining and utilization of other toxic and/or carcinogenic elements such as mercury, lead, and cadmium are likewise spreading them everywhere. Cleanup efforts are so puny in relation to the magnitude of the problem that they are little better than a joke.

The extraction and processing of other resources have followed similar trajectories. Petroleum, long known as a substance that seeped from the ground in places, originally had few uses. But during the 19th century it was discovered that kerosene, distilled from petroleum, could be burned for illumination in lamps, and for that purpose was superior to whale oil. As a result of this discovery the first "oil well" was drilled in Pennsylvania in 1859, and drilling elsewhere soon followed. The petroleum industry at that time was based mainly on kerosene; there was little demand for other petroleum products, such as natural gas and gasoline. But natural gas later came to be used on a large scale for heating, cooking, and illumination, and after the advent of the gasoline-powered automobile around the beginning of the 20th century the petroleum industry won a position of central importance in the economy of the industrialized world. From that time on, new uses for petroleum products have continually been discovered. In addition, processes have been developed for transforming hydrocarbons so that formerly useless petroleum distillates can be turned into useful products, and oil deposits that, because of their undesirable characteristics (e.g., high sulfur content), might not have been worth extracting, can now be made valuable.[61]

Oil companies have come up with ever more sophisticated methods for locating petroleum deposits, and this is one of the reasons why estimates of "known oil reserves" keep increasing. But the estimates also increase because previously inaccessible petroleum is made accessible by new technologies that make it profitable to extract petroleum (including natural gas) from ever more difficult sources. Drillers penetrate deeper and deeper into the Earth's crust, and are even able to drill horizontally; "fracking" (hydraulic fracturing) releases new reserves of oil, and especially gas, from shale rock; techniques are under development for utilizing the vast deposits of methane hydrate found on the ocean floor.[62] As a result of all these technical advances more and more of the Earth's surface is raped by the petroleum industry, and for humans who get in the way it's just tough luck. Fracking, for example, is not a benign technique;[63] among other things, wastewater disposal associated with fracking causes earthquakes.[64]

Anyone who thinks the technological world-system is ever going to stop burning fossil fuels (while any are left) is dreaming.[65] But whether or not the system ever renounces such fuels, other destructive sources of energy will be utilized. Nuclear power-plants generate radioactive waste; no provably safe way of disposing of such waste has yet been identified,[66] and the world's leading self-prop systems aren't even trying very hard to find a permanent home for the accumulating radioactive garbage.[67] Of course, the self-prop systems need energy for the maintenance of their power here and now, whereas radioactive waste represents only a danger for the future and, as we've emphasized, natural selection favors self-prop systems that compete for power in the present with little regard for long-term consequences. So nuclear power-plants continue to be built, while the problem of dealing with their burned-out fuel is largely neglected. In fact, the problem of nuclear waste is on track to become totally unmanageable because, instead of a few of the big, old-style reactors, numerous small ones ("mini-nukes") will soon be built,[68] so that every little town can have its own nuclear power-plant.[69] With the big, old-style reactors at least the radioactive wastes have been concentrated at a relatively small number of sites, but with numerous mini-nukes scattered over the world radioactive wastes will be everywhere. One would have to be extraordinarily naïve, or else gifted with a remarkable capacity for self-deception, to believe that each little two-bit burg is going to handle its nuclear waste responsibly. In practice, much of the radioactive material will escape into the environment.

"Green" energy sources aren't going to wean the system from its dependence on fossil fuels and nuclear power. But even if they did, green energy sources don't look so green when one examines them closely. "There's no free lunch when it comes to meeting our energy needs," says the director of the Natural Resources Defense Council's land program. "To get energy, we need to do things that will have impacts."[70]

The construction of wind farms entails the creation of radioactive waste because, as noted earlier, the lightweight permanent magnets in wind turbines require the rare-earth element neodymium. In addition, wind farms kill numerous birds, which fly into the "propellers" of the turbines.[71] Large numbers of new wind-farms are planned in the U.S., China, and presumably other countries as well,[72] and a likely result will be the extermination of many species of birds. "Shawn Smallwood, a Davis, Calif. ecologist and researcher [said:] 'Just the sheer numbers of turbines we're talking about—we're going to be killing so many raptors until there are no more raptors in my opinion.'"[73] Raptors play an important role in controlling rodent populations, so when the raptors are gone more pesticides will have to be used to kill rodents.

The United States has been developing a military robot called the EATR that relies on green energy inasmuch as it "fuels itself by eating whatever biomass"—a renewable resource—"it finds around it."[74] But you can imagine the devastation that would result from a war fought by armies of robots that gobble for fuel whatever biomass they find. And if the biomass-gobbling technology is ever adapted to civilian use, it will endanger every living thing that can be used to satisfy the system's always ravenous appetite for energy.

But solar energy is harmless, right? Well, not quite, for solar panels compete with biological organisms for the light of the Sun. Let's recall what we pointed out earlier, that the technological system invariably expands until it is using all available energy, and then it demands more. If fossil fuels and nuclear power[75] aren't going to satisfy the system's ever-growing demand for energy, then solar panels will be placed wherever sunlight can be collected. This means, inter alia, that solar panels will progressively invade the habitats of living things, depriving them of sunlight and therefore killing most of them. This is not speculation. Plans "to create huge solar energy plants in the deserts" of the western United States—"prime habitat for threatened plants and animals"[76]—are already being carried out.[77] In 2011 Janine Blaeloch, executive director of the Western Lands Project, predicted: "These [solar energy] plants will introduce a huge amount of damage to our public land and habitat."[78] There is reason to believe that Blaeloch's prediction is beginning to come true.[79] And remember, the system's appetite for energy is insatiable: In all probability the development of solar energy will expand until there is no habitat left for living organisms other than the domesticated crops that the system grows to satisfy its own needs.

But there is much more to be taken into account. Notwithstanding the folly of Ray Kurzweil's fantasies of a future technological utopia, he is absolutely right about some things. He quite correctly points out that in thinking about the future most people make two errors: (i) They "consider the transformations that will result from a single trend [or from several specified trends that are already evident] in today's world as if nothing else will change."[80] And (ii) they "intuitively assume that the current rate of progress will continue for future periods," neglecting the unending *acceleration* of technological development.[81] In order to avoid falling into these errors ourselves, we have to remember that the assaults on the terrestrial environment that are known and observable *now* will not in future be the only ones. Just as the use of petroleum distillates in internal combustion engines was undreamed of before 1860 at the earliest,[82] just as the use of uranium as fuel was undreamed of before the discovery of nuclear fission in 1938–39,[83] just as most uses of the rare earths were undreamed of until recent decades, so there will be future uses of resources, future ways of exploiting the environment, future corners for the technological system to invade that at present are still undreamed of. In attempting to estimate the coming damage to our environment, we can't just project into the future the effects of currently known causes of environmental harm; we have to assume that new causes of environmental harm, which no one today can even imagine, will emerge in the future. Moreover, we have to remember that the growth of technology, and with it the exacerbation of the harm that technology does to our environment, will accelerate ever more rapidly over the coming decades. All this being taken into consideration we have to conclude that, in all probability, little or nothing on our planet will much longer remain free of gross disruption by the technological system.

Most people take our atmosphere for granted, as if Providence had decreed once and for all that air should consist of 78% nitrogen, 21% oxygen, and 1% other gasses. In reality our atmosphere in its present form was created, and is still maintained, through the action of living things.[84] Originally the atmosphere contained far more carbon dioxide than it does today,[85] and we may wonder why the greenhouse effect didn't make the Earth too hot for life ever to begin. The answer, presumably, is that the Sun at that time radiated much less energy than it does now.[86] In any case, it was the biosphere that took the excess carbon dioxide out of the air:

> As primitive bacteria and cyanobacteria had, through photosynthesis or related life processes, captured atmospheric carbon, depositing it on the seafloor, carbon was removed from the atmosphere. ...
>
> Cyanobacteria also were the first organisms to utilize water as a source of electrons and hydrogen in the photosynthetic process. Free oxygen was released as a result of this reaction and began to accumulate in the atmosphere, allowing oxygen-dependent life-forms to evolve.[87]

Biological processes also affect the amount of methane in the atmosphere,[88] and let's remember that methane has a far more powerful effect in promoting global warming than carbon dioxide does.[89] On the other hand, some experts claim that 3.7 billion years ago certain microbes generated large quantities of methane that, instead of warming the planet, *cooled* it by creating clouds that reflected sunlight back into space. Supposedly, the Earth narrowly escaped becoming too cold for the survival of life.[90] However that may be, it's evident that a really radical disruption of the biosphere could cause an atmospheric disaster: a lack of oxygen, a concentration of toxic gasses such as methane or ammonia, a deficiency or an excess of carbon dioxide that would make our planet too cold or too hot to support life.

At present, the most imminent danger seems to be the possible overheating of the Earth through an excess in the atmosphere of carbon dioxide and perhaps methane.[91] Just how hot might the Earth get if humans continue to burn fossil fuels? About 56 million years ago there was a massive increase in the amount of carbon dioxide in our atmosphere, estimated to be roughly equal to the amount that would be added now if humans burned off "all the Earth's reserves of coal, oil, and natural gas."[92] The result was a radical change in the terrestrial environment, including a 9° F (5° C) rise in average temperatures[93] and the flooding of substantial parts of the continents.[94] There weren't any mass extinctions,[95] but this should give us no sense of security about the future of the biosphere, because we can't assume that the effect of adding a given amount of carbon dioxide to the atmosphere today will be the same as what it was 56 million years ago.[96]

The carbon dioxide added to the atmosphere 56 million years ago was probably added relatively slowly, over thousands of years.[97] If humans now burn off all petroleum reserves they undoubtedly will do so in a small fraction of that time, hence living organisms will have little opportunity to adapt to their changed environment. Moreover, the presumed equivalence of the amount of carbon dioxide being released today with what was released 56 million years ago is based on an estimate of the Earth's fossil-fuel reserves that almost certainly is far too low, for new and unexpected deposits of oil and natural gas are continually being discovered and estimates of the reserves are correspondingly raised. Account must also be taken of other ways in which humans add carbon dioxide to the atmosphere. For example, vast quantities of limestone are "burned" to make lime and portland cement[98]: $CaCO_3 \rightarrow CaO + CO_2$. It's not clear how much of the carbon dioxide (CO_2) is eventually recaptured by the lime (CaO) or how long that takes.

But even if the Earth warms no more than it did 56 million years ago, the consequences will be unacceptable to the powerful classes in our society. The world's dominant self-prop systems will therefore resort to "geo-engineering," that is, to a system of artificial manipulation of the atmosphere designed to keep temperatures within acceptable limits.[99] The implementation of geo-engineering will entail immediate, desperate risks,[100] and even if no immediate disaster ensues the eventual consequences very likely will be catastrophic.[101]

Chlorofluorocarbons ("CFCs") have been phased out by international agreement in order to allow the ozone layer, which protects living organisms from the Sun's ultraviolet radiation, to recover from the damage it has suffered in the past. The program has been a clear success,[102] and some people have suggested that the ozone agreement could provide a "template" for an international treaty to limit carbon-dioxide emissions.[103] But the agreement to phase out CFCs was possible only because CFCs are of relatively minor economic importance and substitutes for them can be found.[104] Fossil fuels on the other hand are of central importance in the economies of all industrialized nations and those that are in the process of industrializing; consequently it is safe to say that whatever is done about the greenhouse effect will be too little and too late.

To the greenhouse effect we have to add numerous other factors that tend to disrupt the biosphere. As we've seen, living organisms will be progressively robbed of sunlight by continual expansion of the system's solar-energy installations. There will be no limit to the contamination of our environment with radioactive waste, with toxic elements such as lead, arsenic, mercury, and cadmium,[105] and with a variety of poisonous chemical compounds.[106] There will be oil spills from time to time, since the safety measures taken by the petroleum industry are never quite sufficient,[107] and in some parts of the world the industry doesn't even make any serious effort to prevent spills.[108]

The foregoing effects of the technological system's activities have long been recognized as harmful, but there can be little doubt that many effects not recognized as harmful today will turn out to be harmful tomorrow, for this has often happened in the past.[109] "It has been estimated that the modern sediment loads of the rivers draining into the Atlantic Ocean may be four to five times greater than the prehistoric rates because of the effects of human activity."[110] How, in the long run, will this affect life in the ocean? Does anyone know? Genes from genetically engineered organisms can, and almost certainly will, be passed to wild plants or animals.[111] What will be the ultimate consequences for the biosphere of this "genetic pollution"? No one knows. Even if these and other effects turn out to be harmless when considered separately and individually, all of the "harmless" effects of the system's activities taken together will surely bring about major alterations in the biosphere.

Here we've done no more than scratch the surface. A full assessment of the ways in which the functioning of the technological world-system currently threatens to disrupt the biosphere would require a vast amount of research, and the results would fill several volumes. Will all of these factors add up to a disruption of the biosphere sufficient to prevent it from performing its function in maintaining the present composition of our atmosphere? It's anybody's guess. But that's not all: Let's not forget that the technological system is still in its infancy in comparison with what it will become over the next several decades. At a rapidly accelerating pace and in ways that no one has yet imagined, we can expect the world's leading self-prop systems to find more and more opportunities to exploit, more and more resources to extract, more and more corners to invade, until little or nothing on this planet is left free of technological intervention—intervention that will be carried out in a mad quest for immediate increments of power and without regard to long-term consequences. In the opinion of this writer, there is a strong probability that if the biosphere is not destroyed outright it will at least be rendered incapable of maintaining any reasonable approximation to the present composition of our atmosphere, without which none of the more complex forms of life on this planet will be able to survive.

One plausible outcome might be that the Earth will end up like the planet Venus:

> It has been suggested that the climate of the Earth could be ultimately unstable. Addition of gasses capable of trapping heat could accelerate the release of H_2O and raise the temperature to a point where the oceans would evaporate.... Some believe that such changes may have occurred on Venus.... Venus is a striking example of the importance of the greenhouse effect. Its atmosphere contains a large concentration of CO_2 [= carbon dioxide].... [T]he Venusian surface temperature is much hotter than the Earth's—about 780° K [507° C or 944° F]—in spite of the fact that Venus absorbs less energy from the Sun because of its ubiquitous cloud cover... ."[112]

To sum up the thesis of this part of the present chapter: If the development of the technological world-system is allowed to proceed to its logical conclusion, it will in all probability leave the Earth uninhabitable for all of the more complex forms of life as we know them today. This admittedly remains unproven; it represents the author's personal opinion. But the facts and arguments offered here are enough at least to show that the opinion can be entertained as a plausible hypothesis, and that it would be rash to assume without further proof that the denouement we are facing will be no worse than earlier extinction events in the Earth's history.

What can be taken as a near certainty is that—*if* the development of the technological system is allowed to proceed to its logical conclusion—the outcome for the biosphere will be thoroughly devastating; if it isn't worse than the extinction event at the end of the Cretaceous when the dinosaurs disappeared, it can't be much better; if any humans are left alive, they will be very few; and the technological system itself will be dead.

But note the reservation in the foregoing statement: "*if* the development of the technological system is allowed to proceed to its logical conclusion." The author has occasionally been asked: "If the system is going to destroy itself anyway, then why bother to overthrow it?" The answer, of course, is that if the technological system were eliminated *now* a great deal could still be saved. The longer the system is allowed to continue its development, the worse will be the outcome for the biosphere and for the human race, and the greater will be the risk that the Earth will be left a dead planet.[113]

V. *The techies' wet-dreams.* There is a current of thought that appears to be carrying many technophiles out of the realm of science and into that of science fiction.[114] For convenience, let's refer to those who ride this current as "the techies."[115] The current runs through several channels; not all techies think alike. What they have in common is that they take highly speculative ideas about the future of technology as near certainties, and on that basis predict the arrival within the next few decades of a kind of technological utopia. Some of the techies' fantasies are astonishingly grandiose. For example, Ray Kurzweil believes that "[w]ithin a matter of centuries, human intelligence will have re-engineered and saturated all the matter in the universe."[116] The writing of Kevin Kelly, another techie, is often so vague as to border on the meaningless, but he *seems* to say much the same thing that Kurzweil does about human conquest of the universe: "The universe is mostly empty because it is waiting to be filled with the products of life and the technium... ."[117] "The technium" is Kelly's name for the technological world-system that humans have created here on Earth.[118]

Most versions of the technological utopia include immortality (at least for techies) among their other marvels. The immortality to which the techies believe themselves destined is conceived in any one of three forms:

(i) the indefinite preservation of the living human body as it exists today;[119]

(ii) the merging of humans with machines and the indefinite survival of the resulting man-machine hybrids;[120]

(iii) the "uploading" of minds from human brains into robots or computers, after which the uploaded minds are to live forever within the machines.[121]

Of course, if the technological world-system is going to collapse in the not-too-distant future, as we've argued it must, then no one is going to achieve immortality in any form. But even assuming that we're wrong and that the technological world-system will survive indefinitely, the techies' dream of an unlimited life-span is still illusory. We need not doubt that it will be technically feasible in the future to keep a human body, or a man-machine hybrid, alive indefinitely. It is seriously to be doubted that it will ever be feasible to "upload" a human brain into electronic form with sufficient accuracy so that the uploaded entity can reasonably be regarded as a functioning duplicate of the original brain. Nevertheless, we will assume in what follows that each of the solutions (i), (ii), and (iii) will become technically feasible at some time within the next several decades.

It is an index of the techies' self-deception that they habitually assume that anything they consider desirable will actually be done when it becomes technically feasible. Of course, there are lots of wonderful things that already are and for a long time have been technically feasible, but don't get done. Intelligent people have said again and again: "How easily men could make things much better than they are—if they only all tried together!"[122] But people never do "all try together," because the principle of natural selection guarantees that self-prop systems will act mainly for their own survival and propagation in competition with other self-prop systems, and will not sacrifice competitive advantages for the achievement of philanthropic goals.[123]

Because immortality, as the techies conceive it, will be technically feasible, the techies take it for granted that some system to which they belong can and will keep them alive indefinitely, or provide them with what they need to keep themselves alive. Today it would no doubt be technically feasible to provide everyone in the world with everything that he or she needs in the way of food, clothing, shelter, protection from violence, and what by present standards is considered adequate medical care—if only all of the world's more important self-propagating systems would devote themselves unreservedly to that task. But that never happens, because the self-prop systems are occupied primarily with the endless struggle for power and therefore act philanthropically only when it is to their advantage to do so. That's why billions of people in the world today suffer from malnutrition, or are exposed to violence, or lack what is considered adequate medical care.

In view of all this, it is patently absurd to suppose that the technological world-system is ever going to provide seven billion human beings with everything they need to stay alive indefinitely. If the projected immortality were possible at all, it could only be for some tiny subset of the seven billion—an elite minority. Some techies acknowledge this.[124] One has to suspect that a great many more recognize it but refrain from acknowledging it openly, for it is obviously imprudent to tell the public that immortality will be for an elite minority only and that ordinary people will be left out.

The techies of course assume that they themselves will be included in the elite minority that supposedly will be kept alive indefinitely. What they find convenient to overlook is that self-prop systems, in the long run, will take care of human beings—even members of the elite—only to the extent that it is to the systems' advantage to take care of them. When they are no longer useful to the dominant self-prop systems, humans—elite or not—will be eliminated. In order to survive, humans not only will have to be useful; they will have to be more useful in relation to the cost of maintaining them—in other words, they will have to provide a better cost-versus-benefit balance—than any non-human substitutes. This is a tall order, for humans are far more costly to maintain than machines are.[125]

It will be answered that many self-prop systems—governments, corporations, labor unions, etc.—do take care of numerous individuals who are utterly useless to them: old people, people with severe mental or physical disabilities, even criminals serving life sentences. But this is only because the systems in question still need the services of the majority of people in order to function. Humans have been endowed by evolution with feelings of compassion, because hunting-and-gathering bands thrive best when their members show consideration for one another and help one another.[126] As long as self-prop systems still need people, it would be to the systems' disadvantage to offend the compassionate feelings of the useful majority through ruthless treatment of the useless minority. More important than compassion, however, is the self-interest of human individuals: People would bitterly resent any system to which they belonged if they believed that when they grew old, or if they became disabled, they would be thrown on the trash-heap.

But when *all* people have become useless, self-prop systems will find no advantage in taking care of anyone. The techies themselves insist that machines will soon surpass humans in intelligence.[127] When that happens, people will be superfluous and natural selection will favor systems that eliminate them—if not abruptly, then in a series of stages so that the risk of rebellion will be minimized.

Even though the technological world-system still needs large numbers of people for the present, there are now more superfluous humans than there have been in the past because technology has replaced people in many jobs and is making inroads even into occupations formerly thought to require human intelligence.[128] Consequently, under the pressure of economic competition, the world's dominant self-prop systems are already allowing a certain degree of callousness to creep into their treatment of superfluous individuals. In the United States and Europe, pensions and other benefits for retired, disabled, unemployed, and other unproductive persons are being substantially reduced;[129] at least in the U.S., poverty is increasing;[130] and these facts may well indicate the general trend of the future, though there will doubtless be ups and downs.

It's important to understand that in order to make people superfluous, machines will not have to surpass them in general intelligence but only in certain specialized kinds of intelligence. For example, the machines will not have to create or understand art, music, or literature, they will not need the ability to carry on an intelligent, non-technical conversation (the "Turing test"[131]), they will not have to exercise tact or understand human nature, because these skills will have no application if humans are to be eliminated anyway. To make humans superfluous, the machines will only need to outperform them in making the technical decisions that have to be made for the purpose of promoting the short-term survival and propagation of the dominant self-prop systems. So, even without going as far as the techies themselves do in assuming intelligence on the part of future machines, we still have to conclude that humans will become obsolete. Immortality in the form (i)—the indefinite preservation of the human body as it exists today—is highly improbable.

The techies—or more specifically the transhumanists—will argue that even if the human body and brain as we know them become obsolete, immortality in the form (ii) can still be achieved: Man-machine hybrids will permanently retain their usefulness, because by linking themselves with ever-more-powerful machines human beings (or what is left of them) will be able to remain competitive with pure machines.[132]

But man-machine hybrids will retain a biological component derived from human beings only as long as the human-derived biological component remains useful. When purely artificial components become available that provide a better cost-versus-benefit balance than human-derived biological components do, the latter will be discarded and the man-machine hybrids will lose their human aspect to become wholly artificial.[133] Even if the human-derived biological components are retained they will be purged, step by step, of the human qualities that detract from their usefulness. The self-prop systems to which the man-machine hybrids belong will have no need for such human weaknesses as love, compassion, ethical feelings, esthetic appreciation, or desire for freedom. Human emotions in general will get in the way of the self-prop systems' utilization of the man-machine hybrids, so if the latter are to remain competitive they will have to be altered to remove their human emotions and replace these with other motivating forces. In short, even in the unlikely event that some biological remnants of the human race are preserved in the form of man-machine hybrids, these will be transformed into something totally alien to human beings as we know them today.

The same applies to the hypothesized survival of human minds in "uploaded" form inside machines. The uploaded minds will not be tolerated indefinitely unless they remain useful (that is, more useful than any substitutes not derived from human beings), and in order to remain useful they will have to be transformed until they no longer have anything in common with the human minds that exist today.

Some techies may consider this acceptable. But their dream of immortality is illusory nonetheless. Competition for survival among entities derived from human beings (whether man-machine hybrids, purely artificial entities evolved from such hybrids, or human minds uploaded into machines), as well as competition between human-derived entities and those machines or other entities that are not derived from human beings, will lead to the elimination of all but some minute percentage of all the entities involved. This has nothing to do with any specific traits of human beings or of their machines; it is a general principle of evolution through natural selection. Look at biological evolution: Of all the species that have ever existed on Earth, only some tiny percentage have direct descendants that are still alive today.[134] On the basis of this principle alone, and even discounting everything else we've said in this chapter, the chances that any given techie will survive indefinitely are minute.

The techies may answer that even if almost all biological species are eliminated *eventually*, many species survive for thousands or millions of years, so maybe techies too can survive for thousands or millions of years. But when large, rapid changes occur in the environment of biological species, both the rate of appearance of new species and the rate of extinction of existing species are greatly increased.[135] Technological progress constantly accelerates, and techies like Ray Kurzweil insist that it will soon become virtually explosive;[136] consequently, changes come more and more rapidly, everything happens faster and faster, competition among self-prop systems becomes more and more intense, and as the process gathers speed the losers in the struggle for survival will be eliminated ever more quickly. So, on the basis of the techies' own beliefs about the exponential acceleration of technological development, it's safe to say that the life-expectancies of human-derived entities, such as man-machine hybrids and human minds uploaded into machines, will actually be quite short. The seven-hundred-year or thousand-year life-span to which some techies aspire[137] is nothing but a pipe-dream.

Singularity University, which we discussed in Part VI of Chapter One of this book, purportedly was created to help technophiles "guide research" and "shape the advances" so that technology would "improve society." We pointed out that Singularity University served in practice to promote the interests of technology-orientated businessmen, and we expressed doubt that the majority of technophiles fully believed in the drivel about "shaping the advances" to "improve society." It does seem, however, that *the techies*—the subset of the technophiles that we specified at the beginning of this Part V of the present chapter—are entirely sincere in their belief that organizations like Singularity University[138] will help them to "shape the advances" of technology and keep the technological society on the road to a utopian future. A utopian future will have to exclude the competitive processes that would deprive the techies of their thousand-year life-span. But we showed in Chapter One that the development of our society can never be subject to rational control: The techies won't be able to "shape the advances" of technology, guide the course of technological progress, or exclude the intense competition that will eliminate nearly all techies in short order.

In view of everything we've said up to this point, and in view moreover of the fact that the techies' vision of the future is based on pure speculation and is unsupported by evidence,[139] one has to ask how they can believe in that vision. Some techies, e.g., Kurzweil, do concede a slight degree of uncertainty as to whether their expectations for the future will be realized,[140] but this seems to be no more than a sop that they throw to the skeptics, something they have to concede in order to avoid making themselves too obviously ridiculous in the eyes of rational people. Despite their pro forma admission of uncertainty, it's clear that most techies confidently expect to live for many centuries, if not forever, in a world that will be in some vaguely defined sense a utopia.[141] Thus Kurzweil states flatly: "We will be able to live as long as we want... ."[142] He adds no qualifiers—no "probably," no "if things turn out as expected." His whole book reveals a man intoxicated with a vision of the future in which, as an immortal machine, he will participate in the conquest of the universe. In fact, Kurzweil and other techies are living in a fantasy world.

The techies' belief-system can best be explained as a religious phenomenon,[143] to which we may give the name "Technianity." It's true that Technianity at this point is not strictly speaking a religion, because it has not yet developed anything resembling a uniform body of doctrine; the techies' beliefs are widely varied.[144] In this respect Technianity probably resembles the inceptive stages of many other religions.[145] Nevertheless, Technianity already has the earmarks of an apocalyptic and millenarian cult: In most versions it anticipates a cataclysmic event, the Singularity,[146] which is the point at which technological progress is supposed to become so rapid as to resemble an explosion. This is analogous to the Judgment Day[147] of Christian mythology or the Revolution of Marxist mythology. The cataclysmic event is supposed to be followed by the arrival of techno-utopia (analogous to the Kingdom of God or the Worker's Paradise). Technianity has a favored minority—the Elect—consisting of the techies (equivalent to the True Believers of Christianity or the Proletariat of the Marxists[148]). The Elect of Technianity, like that of Christianity, is destined to Eternal Life; though this element is missing from Marxism.[149]

Historically, millenarian cults have tended to emerge at "times of great social change or crisis."[150] This suggests that the techies' beliefs reflect not a genuine confidence in technology, but rather their own anxieties about the future of the technological society—anxieties from which they try to escape by creating a quasi-religious myth.

NOTES

1. From a speech delivered by Solzhenitsyn in Vaduz, Liechtenstein, Sept. 1993. Quoted by Remnick, p. 21. Here Solzhenitsyn is referring to the famous article by Francis Fukuyama (see List of Works Cited).

2. From "Self Reliance" (1841), in Emerson, p. 30. With this quote we do not mean to express a moral judgment about power in nature or elsewhere, but only an empirical fact about power.

3. See Kaczynski, Letter to David Skrbina: Oct. 12, 2004, Part III. According to Orr, p. 80, "In... 'Darwin's Dangerous Idea,' [Daniel] Dennett proclaimed that natural selection... helps to explain... the twists and turns of human cultural change." I haven't seen Dennett's book and I don't know to what extent, if any, the present chapter parallels or contradicts his work.

4. R. Heilbroner & A. Singer, pp. 26–27.

5. A "nuclear family" is the basic human family consisting of a woman, a man, and any juvenile offspring they may have.

6. When we refer to "competition," we don't necessarily mean intentional or willful competition. Competition, as we use the term, is just something that happens. For example, plants certainly have no intention to compete with one another. It is simply a fact that the plants that most effectively survive and propagate themselves tend to replace those plants that less effectively survive and propagate themselves. "Competition" in this sense of the word is just an inevitable process that goes on with or without any intention on the part of the competitors.

7. Something along these lines, but more complicated, probably happened among the ancient Maya. It's unlikely that the kind of competition we've described here was the sole cause of the collapse of the "Classic" Maya civilization, but it probably was at least a contributing factor and it *may* have been the most important factor. See: Diamond, pp. 157–177, 431. Sharer, pp. 355–57. NEB (2003), Vol. 7, "Maya," p. 970; Vol.15, "Central America," p. 665; Vol. 26, "Pre-Columbian Civilizations," p. 17. "Clean" historical examples are hard to find, because the causes of historical events tend to be complex and open to dispute; the Maya case illustrates this very well. For further discussion, see Appendix Two, Part A.

8. When we refer to the exercise of "foresight" or to the "pursuit" of advantage, our reference is not limited to conscious, intelligent foresight or to intentional pursuit of advantage. We include any behavior (interpreting that word in the broadest possible sense) that has the same effect as the exercise of foresight or the pursuit of advantage, regardless of whether the behavior is guided by any mechanism that could be described as "intelligence." (Compare note 6.) For example, any vertebrates that, in the process of evolving into land animals, had the "foresight" to "attempt" to retain their gills (an advantage if they ever had to return to water) were at a disadvantage due to the biological cost of maintaining organs that were useless on land. Hence they lost out in "competition" with those incipient land animals that "pursued" their short-term advantage by getting rid of their gills. By losing their gills, reptiles, birds, and mammals have become dependent on access to the atmosphere; and that's why whales today will drown if forced to remain submerged too long.

9. The term "prompt" as used here is relative to the circumstances in which the self-prop system exists and the rapidity with which events that are important to it can be expected to occur. A hunting-and-gathering band might keep itself adequately informed about the condition of its territory even if it visited parts of it only once a year. At the other extreme, an advanced technological society needs almost instant long-distance communications.

10. See Appendix Two, Part B.

11. The maximum geographical size of pre-industrial empires was determined not only by factors of transportation and communication, but also by organizational factors such as bureaucratization. However, for any given level of organization, it appears that empires tended to grow to the maximum size

permitted at that level by the existing means of transportation and communication. See Taagepera, pp. 121–23.

12. "Of toxic bonds and crippled nuke plants," *The Week*, Jan. 28, 2011, p. 42 (using the term "tightly linked" in place of "tightly coupled"). Harford, p. 27. See also Perrow, *Normal Accidents*, pp. 89–100; "Black Swans," *The Week*, April 8, 2011, p. 13.

13. Harford, p. 27. *The Week*, April 8, 2011, p. 13.

14. Perrow, *Next Catastrophe*, Chapt. 9. See *The Atlantic*, Jan./Feb. 2015, p. 25, col. 1 (our big banks are "still too interconnected").

15. This argument of course assumes that the most powerful self-prop subsystems will be "intelligent" enough to distinguish between a situation in which their supersystem is subject to an immediate external threat, and a situation in which their supersystem is not subject to such a threat. The assumption, however, will surely be correct in the contexts that are relevant for our purposes.

16. Silverman, p. 103. (Punctuation, capitalization, and so forth have been modernized here for the sake of readability.) Compare Alinsky, p. 149 (the struggle for power among powerful groups "permits only temporary truces, and only when [the powerful groups are] equally confronted by a common enemy").

17. See Appendix Two, Part C.

18. See Appendix Two, Part D.

19. E.g., Anonymous and the now-defunct LulzSec. *The Economist*, June 18, 2011, pp. 67–68; Aug. 6, 2011, pp. 49–50. Saporito, pp. 50–52, 55. Acohido, "Hactivist group." p. 1B, and "LulzSec's gone," p. 1B.

20. E.g., Scandinavian biker gangs apparently have proven very difficult for the authorities to control. *The Week*, Aug. 20, 2010, p. 15. Authorities seem almost helpless against Chinese gangs that produce technologically sophisticated fake IDs that are good enough to fool even experts. *USA Today*, June 11, 2012, p. 1A; Aug. 7, 2012, p. 4A. Cybergangs that use the Internet for criminal purposes are technologically sophisticated and hard to stop. Acohido, "Hackers mine ad strategies," p. 2B. Leger & Arutunyan, pp. 1A, 7A. *USA Today*, Aug. 29, 2013, p. 2B.

21. See notes 66, 70 to Chapter Three. Also: *The Week*, May 21, 2010, p. 8; May 28, 2010, p. 6; Aug. 13, 2010, p. 6; Dec. 24, 2010–Jan. 7, 2011, p. 20. *USA Today*, Nov. 22, 2013, p. 8A.

22. Kenya has been called a "narco-state," *The Week*, Jan. 14, 2011, p. 18, and there is plenty of evidence that this is not far from the truth. Gastrow, Dec. 2011, Chapt. One, especially pp. 24, 26, 28–34. "Available information does not... justify categorizing Kenya as a captured or criminalized state, but the country is clearly on its way to achieving that... status." Gastrow, Sept. 2011, p. 10. The drug gangs involved operate internationally and have massively corrupted the governments of other African countries, such as Guinea-Bissau. O'Regan, p. 6.

23. See Patterson, pp. 9–10, 63.

24. Vick, pp. 46–51. *The Economist*, Dec. 10, 2011, p. 51.

25. R. Heilbroner & A. Singer, pp. 58–60.

26. Ibid., pp. 232–33, 239. Rothkopf, p. 44. Foroohar, "Companies Are The New Countries," p. 21. Corporations are also a dominant force *within* the U.S. political system, because their wealth enables them to offer politicians campaign contributions that in practice function as bribes. See *The Week*, Feb. 25, 2011, p. 16.

27. Carrillo, pp. 77–78. "U.S. Supreme Court Justice William O. Douglas told [President] Franklin Roosevelt that government agencies more than ten years old should be abolished. After that point, they become more concerned with their image than with their mission." David Brower, "Foreword," in Wilkinson, p. ix. See also Keefe, p. 42, quoting Max Weber on bureaucracies' "pure interest... in power."

28. See Carrillo, pp. 207–08.

29. Pakistan: *Time*, May 23, 2011, p. 41. *The Week*, Nov. 26, 2010, p. 15. *The Economist*, Feb. 12, 2011, p. 48; Feb. 26, 2011, p. 65 ("General Ashfaq Kayani... [is] widely seen as the most powerful in [Pakistan]"); April 2, 2011, pp. 38–39; May 21, 2011, p. 50 ("India's most senior security officials say that Pakistan is still, in essence, a state run by its army"); June 18, 2011, p. 47 (calling Pakistan's army "the country's dominant institution"); July 30, 2011, p. 79. *USA Today*, May 13, 2013, p. 5A ("Despite protests over vote-rigging..., observers heralded Pakistan's elections as a historic democratic exercise in a nation known for military takeovers. ... ." But: "Athar Hussain, director of the Asia Research Center at the London School of Economics, said... 'The army will still remain one of the most powerful forces in Pakistan'... .").

Recent (since 2011) events in Egypt have been massively publicized, and it should be obvious to the reader that the army is calling the shots in that country. As an example, we quote *USA Today*, Aug. 16, 2013, p. 1A:

"Egypt's military ousted [Mohammed] Morsi on July 3 [2013] after millions protested Morsi's policies as a new dictatorship of Islamists. ... Egyptian military chief Abdel Fatah al-Sisi has criticized [President] Obama for refusing to endorse the ouster of Morsi... . The Obama administration has not called the ouster a 'military coup'... ."

See also ibid., pp. 5A, 6A, and ibid., Oct. 30, 2013, p. 7A ("In a political vacuum, [Egypt's] top army chief has edge").

30. For the sake of simplicity we define a lineage to be any sequence of organisms $O_1, O_2, O_3, \ldots O_N$ such that O_2 is an offspring of O_1, O_3 is an offspring of O_2, O_4 is an offspring of O_3, and so on down to O_N. We say that such a lineage has survived to the Nth generation. But if O_N produces no offspring, then the lineage does not survive to generation N+1. For example, if John is the son of Mary and George is the son of John and Laura is the daughter of George, then Mary-John-George-Laura is a lineage that survives to the fourth generation. But if Laura produces no offspring, then the lineage does not survive to the fifth generation.

31. See Appendix Two, Part E.

32. Sodhi, Brook & Bradshaw, pp. 515, 517, 519. Benton, p. vii.

33. Kurzweil, pp. 344–49.

34. Ibid., p. 348. Kurzweil refers to an estimate that there should be "billions" of technologically advanced civilizations within the range of our observation, but he plausibly argues that the assumptions on which this estimate is based are highly uncertain and probably overoptimistic (this writer would say wildly overoptimistic). Ibid., pp. 346–47, 357. On the other hand, since Kurzweil wrote in 2005 there have been numerous media reports of discoveries that indicate an abundance of planets, not so far from Earth, on which, as far as anyone can tell, life could have evolved. E.g.: *The Week*, June 3, 2011, p. 21; Sept. 30, 2011, p. 23; Jan. 27, 2012, p. 19. *Time*, June 6, 2011, p. 18. *The Economist*, Dec. 10, 2011, p. 90. *USA Today*, Feb. 7, 2013, p. 5A; April 19–21, 2013, p. 7A; Nov. 5, 2013, p. 5A; May 3, 2016, pp. 1A, 3A; May 11, 2016, p. 8A. Lieberman, pp. 36–39. So an explanation is needed for the fact that our astronomers have detected no indication of *any* extraterrestrial civilizations at *all*. See Kurzweil, p. 357. It should be noted that in this connection Kurzweil egregiously misuses the "anthropic principle." Ibid.

35. From our remarks about Social Darwinism in Part I of this chapter, it should be clear that our intention here is not to exalt competition or portray it as desirable. We aren't making value-judgments in that regard. Our purpose is only to set forth the relevant facts, however unpleasant those facts may be.

36. E.g.: "As [Barbara] Tuchman put it..., 'Chief among the forces affecting political folly is lust for power... .'" Diamond, p. 431.

37. E.g.: "Governments... regularly operate on a short-term focus: they... pay attention only to problems that are on the verge of explosion. For example, a friend of mine who is closely connected to the current [George W. Bush] federal administration in Washington, D.C., told me that, when he visited Washington for the first time after the 2000 national elections, he found that our government's new leaders had what he termed a '90-day focus': they talked only about those problems with the potential to cause a disaster within the next 90 days." Ibid., p. 434.

38. See Appendix Two, Part F.

39. For other parts of the reason, see Kaczynski, Letters to David Skrbina: Aug. 29, 2004, point (I); Nov. 23, 2004, Part IV. E, point 1; March 17, 2005, Part I.A, points 6–8, 10–16, Part II.A, point 3, Part II.B, point 1, Part III.B, points 3–6.

40. This assumption is implicit in, e.g., Benton, pp. vi, viii; McKinney & Lockwood, p. 452; Feeney, pp. 20–21.

41. See Benton, p. vii.

42. Ibid., p. iv. NEB (2007), Vol. 4, "dinosaur," p. 104; Vol. 17, "Dinosaurs," pp. 317–18.

43. See note 42.

44. Duxbury & Duxbury, pp. 111–12, 413–14. Zierenberg, Adams &

Arp. Beatty et al.

45. *The Week*, June 8, 2012, p. 21.

46. Kerr, p. 703.

47. *Popular Science*, June 2013, p. 97.

48. Jonathan Swift, "On Poetry: A Rhapsody," in Browning, p. 274.

49. Zierenberg, Adams & Arp, p. 12962.

50. Kerr, p. 703.

51. Klemm, pp. 147–48.

52. *Evolutionary and Revolutionary Technologies for Mining*, pp. 19–24. See our List of Works Cited—Works Without Named Author.

53. E.g., miners have learned to use cyanide solutions and mercury—both highly poisonous—to leach gold out of sediments or crushed rock. Zimmermann, pp. 270–71, 276. NEB (2002), Vol. 21, "Industries, Extraction and Processing," pp. 491–92. At least in the case of cyanide leaching, this can be done profitably even where only a minute quantity of gold is present in each ton of material treated. Diamond, p. 40. Low-grade copper ores were not utilized until about 1900, when Daniel C. Jackling devised methods that made it possible to mine and process such ores at a profit. *World Book Encyclopedia* (2015), Vol. 4, "Copper," p. 1044. Modern methods of processing copper ores are described in *McGraw-Hill Encyclopedia of Science & Technology* (2012), Vol. 4, "Copper metallurgy," pp. 765–68. Methods have been developed for utilizing low-grade iron ores such as taconite. NEB (2003), Vol. 29, "United States of America," p. 372. See Zimmermann, pp. 271–73. Some iron ores contained too much phosphorus, so that steel produced from them was "almost unfit for practical purposes." Ibid., p. 284. Manchester, p. 32. The utilization of these ores was made possible by the invention at some time between 1875 and 1879 (sources are inconsistent as to the date) of the Thomas-Gilchrist process for making low-phosphorus steel from high-phosphorus ore. Zimmermann, p. 284. NEB (2003), Vol. 5, "Gilchrist, Percy (Carlyle)," p. 265; Vol. 11, "Thomas, Sidney Gilchrist," p. 716; Vol. 21, "Industries, Extraction and Processing," pp. 420, 422, 447–48.

54. E.g., Watson, p. 1A (widespread mercury contamination from old gold-mining operations); Diamond, pp. 36–37, 40–41, 453–57.

55. Diamond, pp. 455–56.

56. Folger, pp. 138, 140, notes the current indispensability of rare earths; NEB (2007), Vol. 15, "Chemical Elements," pp. 1016–17, notes the former limited utility of rare earths. For a detailed description of the vast growth in applications of the rare earths, see Krishnamurthy & Gupta, pp. 33–73. Ibid., p. 73, states: "Over the years, analyzing world rare-earth demand on an annual basis has shown that it has remained more or less the same." This may be true for some limited span of years, say, perhaps, the ten years or so preceding the publication of Krishnamurthy & Gupta's book, but, given the vast expansion in the applications of rare earths,

the statement would be implausible if applied over the long term. Krishnamurthy & Gupta themselves refer on pp. 743–44 to the "continued increase in global usage" and the "fast-expanding world demand" for at least some rare earths. Even if the demand has been static for a few years, it seems unlikely that it will long remain so.

57. Margonelli, p. 17. Folger, loc. cit. (hundreds of pounds of neodymium for a single wind turbine). Krishnamurthy & Gupta, pp. 50–51, provide some technical details.

58. Margonelli, p. 18. Folger, p. 145. Krishnamurthy & Gupta, e.g., p. 718.

59. Ibid., Chapt. 9, pp. 717–744.

60. The Bouma postyard near Lincoln, Montana, which treated posts and poles with cupric arsenate, was in operation throughout the author's 25-year residence in that area.

61. For this whole paragraph see Zimmermann, pp. 323–24, 401–07; NEB (2002), Vol. 21, "Industries, Extraction and Processing," pp. 515, 520, 523–28; Krauss, p. B8; C. Jones, p. 3B. Allan Nevins's biography of John D. Rockefeller (see List of Works Cited), who created the Standard Oil Company, is also of interest in this connection.

62. For this paragraph up to this point, see NEB (2002), Vol. 21, "Industries, Extraction and Processing," pp. 515–19; Mann, pp. 48–63; Walsh, "Power Surge," pp. 36–39; Reed, p. B6; Rosenthal, p. B6; K. Johnson & R. Gold, pp. A1, A6; Vara, pp. 20–21; *USA Today*, May 10, 2011, p. 2A, Nov. 23, 2012, p. 10A, Nov. 4, 2013, p. 3B, and Nov. 14, 2013, p. 1A.

63. See, e.g., Walsh, "Gas Dilemma," pp. 43, 45–46, 48; *USA Today*, July 19, 2016, p. 6B.

64. *The Week*, April 8, 2016, p. 7. *USA Today*, Aug. 11, 2016, p. 4A and Dec. 7, 2016, p. 6B.

65. This conclusion is strongly suggested by the theory of natural selection as developed in the present chapter, and it is supported empirically by the system's failure to solve other problems that require worldwide international cooperation and renunciation of competitive advantages (e.g., the failure to eliminate war or nuclear weapons), as well as the failure to deal with the greenhouse effect itself. Note failure of global-warming summits in Copenhagen, *USA Today*, Nov. 16, 2009, p. 5A and Cancún, *The Week*, Dec. 10, 2010, p. 23, "Climate change: Resignation sets in." The famous "Paris Climate Agreement" was touted as a "turning point for the planet," *USA Today*, Oct. 6, 2016, p. 1A, but President Trump, as we all know, has withdrawn the U.S. from that agreement, and even if the agreement had remained intact it would have accomplished very little toward bringing global warming under control, Lomborg, p. 7A.

66. See note 27 to Chapter One; Wald, "Nuclear Industry Seeks Interim Site," pp. A1, A20, and "What Now for Nuclear Waste?," pp. 48–53.

67. See, e.g., "Radioactive fuel rods: The silent threat," *The Week*, April 15, 2011, p. 13. Even where cleanup efforts are undertaken, they are likely to be characterized by incompetence and inefficiency. See, e.g., *USA Today*, Aug. 29, 2012, p. 2A; May 10, 2017, p. 3A ("Tunnel containing nuclear waste collapses"); June 26, 2017, pp. 1A & 2A.

68. Carroll, pp. 30–33. Koch, p. 4B.

69. Carroll, p. 33 ("The isolated Alaska village of Galena is in discussions with Toshiba" to buy a mini-nuke).

70. Matheny, p. 3A.

71. Welch, p. 3A. *The Week*, March 23, 2012, p. 14.

72. Welch, p. 3A. MacLeod, p. 7A.

73. Welch, p. 3A.

74. *The Economist*, April 2, 2011, p. 65.

75. Nuclear energy will include electricity from fusion power-plants if such plants ever become a practical alternative. But as of March 2017 all fusion reactors have consumed more energy than they have produced, and moreover such reactors are very expensive to build. See H. Fountain in our List of Works Cited. So it will be a long time before fusion power-plants become economically viable, if they ever do. Controlled fusion has been touted as an unlimited and perfectly clean source of energy, but in reality fusion power-plants will routinely release some radioactive tritium gas into the atmosphere and will produce radioactive waste that will have to be disposed of. In addition, as with present-day fission power-plants, there will be a possibility of radiation-releasing accidents. See Taylor et al. Even if fusion plants were perfectly clean and economically competitive, we could expect the system's consumption of energy to increase exponentially until some limit were reached. If nothing else, the amount of heat generated would eventually lead—independently of any greenhouse effect—to an intolerable level of global warming.

76. Matheny, p. 3A. See also Lovich & Ennen.

77. See Hernandez et al.; Walsh, "Power Surge," pp. 34–35.

78. Matheny, p. 3A. At this point the first edition of the present work cited an item from *The Week* for the "fact" that the manufacture of solar panels required rare-earth elements, but it now appears that the "fact" is a myth.

79. See Hernandez et al. Also, solar energy plants kill numerous birds. Walston et al. Of course, fossil-fuel power-plants too kill numerous birds, ibid., in addition to all the other environmental damage that they do. Our purpose here is not to show that "green" energy is no better than fossil-fuel energy. Our point is merely that the production of energy even from "green" sources does substantial damage to the environment. Since the technological system's appetite for energy is insatiable, the exploitation even of "green" energy sources will expand without limit and in the long run will devastate our environment just as surely as the use of fossil fuels will.

80. Kurzweil, p. 13. In some important ways Kurzweil himself falls into this error.

81. Ibid., p. 12.

82. According to Zimmermann, p. 323, the first functioning internal combustion engine (fueled by gas) was built in 1860. Internal combustion engines using gasoline and kerosene came later.

83. NEB (2003), Vol. 29, "War, Technology of," p. 575.

84. NEB (2003), Vol. 14, "Atmosphere," pp. 317, 321–22, 330–31, and "Biosphere," p. 1155. Ward, especially pp. 46–53, 75. *World Book Encyclopedia* (2015), Vol. 6, "Earth," p. 26 (the carbon cycle).

85. NEB (2003), Vol. 14, "Biosphere," p. 1155, says that the Earth's atmosphere once was "largely composed of carbon dioxide," but this is unlikely, since ibid., "Atmosphere," p. 321, refers to an "approximately hundredfold decline of atmospheric CO_2 [= carbon dioxide] abundances from [3.5 billion] years ago to the present." The present atmosphere contains roughly 400 parts per million, or 0.04%, of CO_2. Kunzig, p. 96 (chart). So the atmosphere of 3.5 billion years ago must have contained something like 100×0.04%=4% of CO_2. On the other hand, Ward, p. 104, suggests that at that time as much as a third of the Earth's atmosphere may have been CO_2.

86. Estimates of the energy radiated by the Sun 3.5 billion years ago are inconsistent. Compare: NEB (2003), Vol. 14, "Biosphere," p. 1155; ibid., Vol. 27, "Solar System," p. 457; ibid., Vol. 28, "Stars and Star Clusters," p. 199; Ward, pp. 43, 74; Ribas, p. 2. But it seems safe to say that the Sun today radiates somewhere between 25% and 45% more energy than it did 3.5 billion years ago.

87. NEB (2003), Vol. 14, "Biosphere," p. 1155. See also ibid., "Atmosphere," p. 330; Ward, p. 75.

88. NEB (2003), Vol. 14, "Atmosphere," p. 321. Mann, p. 56.

89. E.g., Mann, p. 62.

90. Ward, pp. 74–75. For some remarks on Ward's book, see Appendix Four.

91. Regarding methane see, e.g., *USA Today*, March 5, 2010, p. 3A ("Methane... appears to be seeping through the Arctic Ocean floor and into the Earth's atmosphere..."); Mann, pp. 56, 62.

92. Kunzig, p. 94.

93. Ibid., p. 96 (chart caption).

94. Ibid., pp. 90–91.

95. Ibid., p. 94.

96. Ibid., p. 109 ("That episode doesn't tell us what will happen to life on Earth if we... burn the rest [of our planet's fossil-fuel reserves].").

97. Ibid., pp. 105–08.

98. It is estimated that the manufacture of Portland cement accounts for about five percent of all human-caused emissions of carbon dioxide. *National*

Geographic, Jan. 2016, "Towering Above," unnumbered page.

99. See Wood, pp. 70–76; Sarewitz & Pielke, p. 59; *Time*, March 24, 2008, p. 50.

100. Wood, pp. 72, 73, 76.

101. See Appendix Four.

102. *USA Today*, July 1–4, 2016, p. 1A.

103. See ibid., Sept. 24, 2014, p. 10A.

104. Ibid.

105. E.g., *Science News*, Vol. 163, Feb. 1, 2003, p. 72 (mercury); Batra (cadmium); *USA Today*, Aug. 7, 2014, p. 2A (mercury); ibid., Jan. 20, 2016, p. 8A (lead, but see also ibid., Jan. 27, 2016, p. 7A); *The Week*, Jan. 20, 2012, p. 18 (depleted uranium scattered by non-nuclear artillery shells, which causes birth defects).

106. See notes 20, 21 to Chapter One and, e.g., *Vegetarian Times*, May 2004, p. 13 (quoting *Los Angeles Times* of Jan. 13, 2004); *U.S. News &World Report*, Jan. 24, 2000, pp. 30–31. On cyanide, see notes 53 & 54, above.

107. Regarding the effects of the 2010 oil spill in the Gulf of Mexico, see *Time*, Sept. 27, 2010, p. 18; *The Week*, Sept. 24, 2010, p. 7.

108. *The Week*, June 18, 2010, p.12. Searcy, pp. A4, A6.

109. Examples: Artificial lighting is thought to be partly responsible for dramatic declines in firefly populations. *National Geographic*, June 2009, "ENVIRONMENT: Dimming Lights," unnumbered page. Many thousands of untested chemicals are getting into our environment, *The Week*, March 12, 2010, p. 14 and Dec. 2, 2011, p. 18; *Time*, April 12, 2010, pp. 59–60, and these sometimes turn out to have unexpected harmful effects, e.g., "Shrimp on Prozac," *The Week*, Aug. 6, 2010, p. 19. Exotic species brought into a region in the belief that they are harmless often reproduce uncontrollably and do enormous damage. See note 36 to Appendix Two. The use of plastics has led to serious, totally unforeseen harm to life in the oceans. Duxbury & Duxbury, p. 302. Gardner, Prugh & Renner, pp. 86–87. *USA Today*, March 23, 2018, p. 3A ("Ocean garbage dump...").

110. NEB (2003), Vol. 26, "Rivers." p. 860.

111. E.g., *Denver Post*, Aug. 23, 2005, p. 2B.

112. NEB (2003), Vol. 14, "Atmosphere," p. 331.

113. For some remarks concerning small islands in relation to the theory developed in the present chapter, see Appendix Two, Part G.

114. It is significant that Ray Kurzweil, the best-known of the techie prophets, started out as a science-fiction enthusiast. Kurzweil, p. 1. Kim Eric Drexler, the prophet of nanotechnology, started out "specializing in theories of space travel and space colonization." Keiper, p. 20.

115. The techies of course include the transhumanists, but some techies—as we use the term—do not appear to be transhumanists.

116. Grossman, p. 49, col. 2. Kurzweil, pp. 351–368.

117. Kelly, p. 357.

118. Ibid., pp. 11–12.

119. Grossman, p. 47. Kurzweil, p. 320.

120. Grossman, p. 44, col. 3. Kurzweil, pp. 194–95, 309, 377. Vance, p. 1, col. 3; p. 6, col. 1.

121. Grossman, p. 44, col. 3; p. 48, col. 1; p. 49, col. 1. Kurzweil, pp. 198–203, 325–26, 377. The techies—or more specifically the transhumanists—seem to assume that their own consciousness will survive the uploading process. On that subject Kurzweil is somewhat equivocal, but in the end seems to assume that his consciousness will survive if his brain is replaced with nonbiological components not all at once, but bit by bit over a period of time. Kurzweil, pp. 383–86.

122. Winston Churchill, Sept. 15, 1909, quoted by Jenkins, p. 212. Other examples: "... liberty, toleration, equality of opportunity, socialism... there is no reason why any of them should not be fully realised, in a society or in the world, if it were the united purpose of a society or of the world to realise it." Bury, p. 1 (originally published in 1920; see ibid., p. xvi). On July 22, 1944, John Maynard Keynes noted that forty-four nations had been learning to "work together." He added: "If we can so continue... [t]he brotherhood of man will have become more than a phrase." (Fat chance!) Skidelsky, p. 355.

123. This of course does not mean that no self-prop system *ever* does anything beneficent that is contrary to its own interest, but the occasional exceptions are relatively insignificant. Bear in mind that many apparently beneficent actions are actually to the advantage of the self-prop system that carries them out.

124. Grossman, p. 48, col. 3 ("Who decides who gets to be immortal?"). Vance, p. 6, col. 1.

125. Humans need to be fed, clothed, housed, educated, entertained, disciplined, and provided with medical care. Whereas machines can work continuously with only occasional down-time for repairs, humans need to spend a great deal of time sleeping and resting.

126. Also, modern societies find it advantageous to encourage people's compassionate feelings through propaganda. See Kaczynski, "The System's Neatest Trick," Part 4.

127. Grossman, pp. 44–46. Kurzweil, pp. 135ff and passim. Machines that surpass humans in intelligence might not be digital computers as we know them today. They might have to depend on quantum-theoretic phenomena, or they might have to make use of complex molecules as biological systems do. Grossman, p. 48, col. 2; Kurzweil, pp. 111–122; *USA Today*, March 8, 2017, p. 5B (IBM & other companies are working to develop computers that make use of quantum-theoretic phenomena). This writer has little doubt that, with commitment of sufficient resources over a sufficient period of time, it would be technically feasible to develop artificial devices having general intelligence that surpasses that of humans

("strong artificial intelligence," or "strong AI," Kurzweil, p. 260). See Kaczynski, Letter to David Skrbina: April 5, 2005, first two paragraphs. Whether it would be technically feasible to develop strong AI as soon as Kurzweil, p. 262, predicts is another matter. Moreover, it is seriously to be doubted whether the world's leading self-prop systems will ever have any need for strong AI. If they don't, then there's no reason to assume that they will commit to it sufficient resources for its development. See Somers, pp. 93–94. Contra: *The Atlantic*, July/Aug. 2013, pp. 40–41; *The Week*, Nov. 4, 2011, p. 18. However, the assumption that strong AI will soon appear plays an important role in Kurzweil's vision of the future, so we could accept that assumption and proceed to debunk Kurzweil's vision by reductio ad absurdum. But the argument of Part V of this chapter does not require the assumption that strong AI will ever exist.

128. E.g.: *The Week*, Sept. 30, 2011, p. 14 ("Capitalism is killing the middle class"); Feb. 17, 2012, p. 42 ("No reason to favor manufacturing"); April 6, 2012, p. 11; May 4, 2012, p. 39 ("The half-life of software engineers"); Jan. 29, 2016, p. 32. *USA Today*, July 9, 2010, pp. 1B–2B (machines as stock-market traders); April 24, 2012, p. 3A (computer scoring of essays); Sept. 14, 2012, p. 4F; May 20, 2014, pp. 1A–2A; July 28, 2014, p. 6A; Oct. 29, 2014, pp. 1A, 9A; Feb. 11, 2015, p. 3B; Dec. 22, 2015, p. 1B; Feb. 21, 2017, p. 3B. *The Economist*, Sept. 10, 2011, p. 11 and "Special report: The future of jobs"; Nov. 19, 2011, p. 84. *The Atlantic*, June 2013, pp. 18–20. *Wall Street Journal*, June 13, 2013, p. B6. Davidson, pp. 60–70. Carr, pp. 78–80. Foroohar, "What Happened to Upward Mobility?," pp. 29–30, 34. Markoff, "Skilled Work Without the Worker," pp. A1, A19. Lohr, p. B3. Rotman (entire article). Robots can even perform functions formerly thought to require a "human touch," e.g., they can serve as companions with which people connect emotionally just as they connect with other people. *Popular Science*, June 2013, p. 28. *The Atlantic*, Jan./Feb. 2016, p. 31; March 2017, p. 29.

129. E.g.: *USA Today*, July 20, 2011, p. 3A ("Painful plan in R.I."); Sept. 29, 2011, pp. 1A, 4A; Oct. 24, 2011, p. 1A; Sept. 14, 2012, p. 5A (Spain); Sept. 24, 2012, p. 6B (several European countries); Sept. 28, 2012, p. 5B (Spain); Aug. 5, 2013, p. 3A; Oct. 16–18, 2015, p. 1A; April 26, 2017, pp. 1A–2A. *The Economist*, June 11, 2011, p. 58 (Sweden). *The Week*, April 6, 2012, p. 14 (Greece, Spain); July 29, 2011, p. 12 ("The end of the age of entitlements"). Drehle, p. 32. Sharkey, pp. 36–38. A friend of the author wrote on Oct. 3, 2012: "[My parents] don't have any set up for long term care... and at this point many states... are doing what is called estate recovery and the like, which means that if Dad were to go in a nursing home... either his Veteran's stipend, social security, and pension would all go into paying for the care, meaning Mom would not have enough to live on... or, in a different scenario, Medicaid would put a lien on their house and when he dies, mom would be out of luck so Medicaid could be repaid for his 'care'—which at that low level is very poor care, by selling the house." In regard to probable future

treatment of people who seek immortality: "The frozen head of baseball legend Ted Williams has not been treated well. . . . [A]t one point Williams's head, which the slugger ordered frozen in hopes of one day being brought back to life, was propped up by an empty tuna-fish can and became stuck to it. To detach the can... staff whacked it repeatedly with a monkey wrench, sending 'tiny pieces of frozen head' flying around the room." *The Week*, Oct. 16, 2009, p. 14.

130. E.g.: *USA Today*, Sept. 29, 2011, pp. 1A–2A; Sept. 12, 2016, p. 3A. *The Week*, Sept. 30, 2011, p. 21 ("Poverty: Decades of progress, slipping away"); July 27, 2012, p. 16 ("Why the poor are getting poorer"). Kiviat, pp. 35–37. Also: "Half of all U.S. workers earned less than $26,364 in 2010—the lowest median wage since 1999, adjusted for inflation." *The Week*, Nov. 4, 2011, p. 18. "The average American family's net worth dropped almost 40 percent... between 2007 and 2010." Ibid., June 22, 2012, p. 34. *USA Today*, Sept. 14, 2016, p. 1A, reports: "Household incomes see first big gain since 2007." This no doubt reflects the current (up to Jan. 2018) high point in the economic cycle. As the economic cycle approaches the next low point, incomes likely will decline again.

131. NEB (2003), Vol. 12, "Turing test," p. 56. NEB is more accurate on the Turing test than is Kurzweil, p. 294: In order to pass the test, machines may not have to "emulate the flexibility, subtlety, and suppleness of human intelligence." See, e.g., *The Week*, Nov. 4, 2011, p. 18.

132. Grossman, p. 44, col. 3. Vance, p. 6, col. 4. Kurzweil, pp. 24–25, 309, 377. Man-machine hybrids are also called "cyborgs."

133. Kurzweil, p. 202, seems to agree.

134. "Species come and go continually—around 99.9 per cent [of] all those that have ever existed are now extinct." Benton, p. ii. We assume this means that 99.9 percent have become extinct without leaving any direct descendants that are alive today. Independently of that assumption, it's clear from the general pattern of evolution that only some minute percentage of all species that have ever existed can have descendants that are alive today. See, e.g., NEB (2003), Vol. 14, "Biosphere," pp. 1154–59; Vol. 19, "Fishes," p. 198, and "Geochronology," especially pp. 750–52, 785, 792, 794–95, 797, 802, 813–14, 819, 820, 825–27, 831–32, 836, 838–39, 848–49, 858–59, 866–67, 872. Extinctions have by no means been limited to a few major "extinction events"; they have occurred continually throughout the evolutionary process, though at a rate that has varied widely over time. See Benton, p. ii; NEB (2003), Vol. 18, "Evolution, Theory of," pp. 878–79; NEB (2007), Vol. 17, "Dinosaurs," p. 318.

135. We don't have explicit authority for this statement, though it receives some support from Sodhi, Brook & Bradshaw, p. 518. We make the statement mainly because it's just common sense and seems generally consistent with the facts of evolution. We're betting that most evolutionary biologists would agree with it, though they might add various reservations and qualifications.

136. Grossman, pp. 44–46, 49. Vance, p. 6, cols. 3–5. Kurzweil, e.g., pp. 9, 25 ("an hour would result in a century of progress").

137. Vance, p. 7, col. 1 (700 years). "Mr. Immortality," *The Week*, Nov. 16, 2007, pp. 52–53 (1,000 years).

138. Other such organizations are the Foresight Institute, Keiper, p. 29; Kurzweil, pp. 229, 395, 411, 418–19, and the Singularity Institute, Grossman, p. 48, col. 3; Kurzweil, p. 599n45.

139. There is of course evidence to support many of the techies' beliefs about particular technological developments, e.g., their belief that the power of computers will increase at an ever-accelerating rate, or that it will some day be technically feasible to keep a human body alive indefinitely. But there is no evidence to support the techies' beliefs about the future of society, e.g., their belief that our society will actually keep some people alive for hundreds of years, or will be motivated to expand over the entire universe.

140. Grossman, p. 48, col. 3; p. 49, col. 1 ("the future beyond the Singularity is not knowable"). Vance, p. 7, col. 4. See Kurzweil, pp. 420, 424.

141. "[S]ome people see the future of computing as a kind of heaven." Christian, p. 68. The utopian cast of techie beliefs is reflected in the name of Keiper's journal, *The New Atlantis*, evidently borrowed from the title of an incomplete sketch of a technological "ideal state" that Francis Bacon wrote in 1623. Bury, pp. 59–60&n1. Probably most techies would deny that they are anticipating a utopia, but that doesn't make their vision less utopian. For example, Kelly, p. 358, writes: "The technium... is not utopia." But on the very next page he launches into a utopian rhapsody: "The technium... expands life's fundamental goodness. ... The technium... expands the mind's fundamental goodness. Technology... will populate the world with all conceivable ways of comprehending the infinite." Etc. Kelly's book as a whole can best be described as a declaration of faith.

142. Kurzweil, p. 9.

143. Several observers have noticed the religious quality of the techies' beliefs. Grossman, p. 48, col. 1. Vance, p. 1, col. 4. Markoff, "Ay Robot!," p. 4, col. 2 (columns occupied by advertisements are not counted). Keiper, p. 24. Kurzweil, p. 370, acknowledges the comment of one such observer, then shrugs it off by remarking, "I did not come to my perspective as a result of searching for an alternative to customary faith." But this is irrelevant. St. Paul, according to the biblical account, was not searching for a new faith when he experienced the most famous of all conversions; in fact, he had been energetically persecuting Christians right up to the moment when Jesus allegedly spoke to him. Acts 9: 1–31. Saul = Paul, Acts 13: 9. Certainly many, perhaps the majority, of those who undergo a religious conversion do so not because they have consciously searched for one, but because it has simply come to them.

Like Kurzweil, many techies stand to profit financially from Technianity, but it is entirely possible to hold a religious belief quite sincerely even while one profits from it. See, e.g., *The Economist*, Oct. 29, 2011, pp. 71–72.

144. E.g., Grossman, p. 46, col. 2.

145. Christianity in its inceptive stages lacked a uniform body of doctrine, and Christian beliefs were widely varied. Freeman, passim, e.g., pp. xiii–xiv, 109–110, 119, 141, 146.

146. Grossman, pp. 44–46. Kurzweil, p. 9. Another version of the Singularity is the "assembler breakthrough" posited by nanotechnology buffs. Keiper, pp. 23–24.

147. It's not entirely clear whether the Day of Judgment and the Second Coming of Jesus are supposed to occur at the same time or are to be separated by a thousand years. Compare Relevation 20: 1–7, 12–13 with NEB (2003), Vol. 17, "Doctrines and Dogmas, Religious," p. 406 (referring to "the Second Coming… of Christ… to judge the living and the dead") and ibid., Vol. 7, "Last Judgment," p. 175. But for our purposes this is of little importance.

148. A correspondent (perhaps under the mistaken impression that the proletariat included all of the "lower" classes) has raised the objection that the proletariat was not a minority. Marxist literature is not consistent as to who belongs to the proletariat. For instance, Lenin in 1899 held that the poor peasants constituted a "rural proletariat." See "The Development of Capitalism in Russia," e.g., Conclusions to Chapter II, section 5; in Christman, p. 19. But in 1917 Lenin clearly implied that the peasantry, including the poor peasants, did not belong to the proletariat, which he now identified as "the armed vanguard of all the exploited, of all the toilers." See "The State and Revolution," Chapt. II, section 1; Chapt. III, sections 1 & 3; respectively pp. 287–88, 299, 307 in Christman. It is the proletariat in this sense—the vanguard of all the toilers—that we have in mind when we speak of the Elect of Marxist mythology, and it's clear from Marxist theory generally that the proletariat in this sense was to consist mainly if not exclusively of industrial workers. E.g., Lenin wrote in 1902: "the strength of the modern [socialist] movement lies in the awakening of the masses (principally the *industrial* proletariat)…" (emphasis added). "What is to be Done?," Chapt. II, first paragraph; in Christman, pp. 72–73. Stalin, *History of the Communist Party*, likewise made clear that the proletariat consisted of *industrial* workers and that these at the time of the 1917 revolution comprised only a minority of the population; e.g., first chapter, Section 2, pp. 18, 22; third chapter, Section 3, pp. 104–05 and Section 6, p. 126; fifth chapter, Section 1, p. 201 and Section 2, p. 211. Almost certainly, industrial workers have never constituted a majority of the population of any large country.

149. On the subject of apocalyptic and millenarian cults, see NEB (2003), Vol. 1, "apocalyptic literature" and "apocalypticism," p. 482; Vol. 17, "Doctrines and Dogmas, Religious," pp. 402, 406, 408. Also the Bible, Revelation 20.

150. NEB (2003), Vol. 8, "millennium," p. 133. See also Vol. 17, "Doctrines and Dogmas, Religious," p. 401 ("Eschatological themes thrive particularly in crisis situations..."). See Freeman, p. 15. For millenarian cults in China, see Ebrey, pp. 71, 73, 190, 240; Mote, pp. 502, 518, 520, 529, 533.

CHAPTER THREE

How to Transform a Society: Errors to Avoid

> In studying any complex process in which there are two or more contradictions, we must devote every effort to finding its principal contradiction. Once this principal contradiction is grasped, all problems can be readily solved.
>
> — Mao Zedong[1]

> A proposition must be plain, to be adopted by the understanding of the people. A false notion which is clear and precise will always have more power in the world than a true principle which is obscure or involved.
>
> — Alexis de Tocqueville[2]

In this chapter we will state some rules that deserve the attention of anyone who wants to bring about radical changes in a society. Not all of the rules are precise enough to be easily applied and some may not be applicable in every situation, but if a radical movement fails to take the rules into account it risks throwing away its chances of success.

In the first part of this chapter we will give a brief and simplified explanation of the rules. Further on we will examine the meaning of the rules, illustrate them with examples, and discuss the limits of their applicability. In the last part of the chapter we will show how ignorance of the rules ensures the failure of present-day efforts to deal with the problems generated by modern technology, including the problem of environmental devastation.

I. Postulates and Rules

We begin by stating four postulates. We postpone a discussion of the extent to which the postulates are true.

Postulate 1. You can't change a society by pursuing goals that are vague or abstract. You have to have a clear and concrete goal. As an experienced activist put it: "Vague, over-generalized objectives are seldom met. The trick is to conceive of some specific development which will inevitably propel your community in the direction you want it to go."[3]

Postulate 2. Preaching alone—the mere advocacy of ideas—cannot bring about important, long-lasting changes in the behavior of human beings, unless in a very small minority.[4]

Postulate 3. Any radical movement tends to attract many people who may be sincere, but whose goals are only loosely related to the goals of the movement.[5] The result is that the movement's original goals may become blurred, if not completely perverted.[6]

Postulate 4. Every radical movement that acquires great power becomes corrupt, at the latest, when its original leaders (meaning those who joined the movement while it was still relatively weak) are all dead or politically inactive. In saying that a movement becomes corrupt, we mean that its members, and especially its leaders, primarily seek personal advantages (such as money, security, social status, powerful offices, or a career) rather than dedicating themselves sincerely to the ideals of the movement.

From these postulates we can infer certain rules to which every radical movement should pay close attention.

Rule (i) In order to change a society in a specified way, a movement should select a single, clear, simple, and concrete objective the achievement of which will produce the desired change.

It follows from Postulate 1 that the movement's objectives must be clear and concrete. According to Postulate 3 there will be a tendency for the movement's objectives to become blurred or perverted, and this tendency will be most easily resisted if the movement has only a single objective that is simple in addition to being clear and concrete. As seen in the epigraph, above, Mao emphasized the importance of identifying the "principal contradiction" in any situation, and this one principal contradiction commonly will point to a single, decisive objective that a movement needs to achieve in order to transform a society.

In any conflict situation in which victory is uncertain, it is always essential to concentrate one's efforts on the achievement of the single most critical objective. Military practitioners and theorists like Napoleon and Clausewitz recognized the importance of concentrating one's forces at the decisive point,[7] and Lenin noted that this principle applies in politics as it does in war.[8] But we shouldn't need Napoleon, Clausewitz, or Lenin to tell us this—it's just common sense: When you're facing a difficult struggle and have no strength to spare, you'd better concentrate what strength you have where it will do the most good: on the single most critical objective.

Rule (ii) If a movement aims to transform a society, then the objective selected by the movement must be of such a nature that, once the objective has been achieved, its consequences will be irreversible. This means that, once society has been transformed through the achievement of the objective, society will remain in its transformed condition without any further effort on the part of the movement or anyone else.

In order to transform society, the movement will have to acquire great power and therefore, according to Postulate 4, will soon become corrupt. Once corrupted, the members of the movement or their successors will no longer exert themselves to maintain the transformed condition of society that corresponds to the ideals of the movement, but will be concerned only to gain and hold personal advantages. Consequently, society will not remain in its transformed condition unless the transformation is irreversible.

Rule (iii) Once an objective has been selected, it is necessary to persuade some small minority to commit itself to the achievement of the objective by means more potent than mere preaching or advocacy of ideas. In other words, the minority will have to organize itself for practical action.

As pointed out in Postulate 2, the advocacy of ideas alone cannot change society, so some group will have to be organized for the purpose of applying methods more potent than mere advocacy of ideas. At least at the outset, this group will ordinarily include only a very small minority because, again by Postulate 2, prior to the application of methods more potent than the mere advocacy of ideas, only a very small minority can be persuaded to act.

Rule (iv) In order to keep itself faithful to its objective, a radical movement should devise means of excluding from its ranks all unsuitable persons who may seek to join it.

This can be important, because according to Postulate 3 the admission of unsuitable persons will promote the blurring or perversion of the movement's objective.

Rule (v) Once a revolutionary movement has become powerful enough to achieve its objective, it must achieve its objective as soon as possible, and in any case before the original revolutionaries (meaning those who joined the movement while it was still relatively weak) die or become politically inactive.

As noted earlier, the movement will have to become very powerful in order to achieve its objective, therefore, by Postulate 4, it will soon be corrupted. Once corrupted, the movement will no longer be faithful to its objective, so if the objective is to be achieved at all it must be achieved before the movement becomes corrupt.

II. Examination of the Postulates

Let's take a careful look at the postulates and ask ourselves to what extent they are true.

Postulate 1. To see the truth of this postulate, we don't need to rely on the opinion of the experienced activist quoted above. It should be obvious that vague or abstract goals can't ordinarily serve as a basis for effective action.

For example, "freedom" by itself will not serve as a goal, because different people have different conceptions of what constitutes freedom and of the relative importance of different aspects of freedom. Consequently, effective and consistent cooperation in pursuit of an unspecified "freedom" is impossible. The same is true of other vague goals like "equality," "justice" or "protecting the environment." For effective cooperation you need a clear and concrete goal, so that everyone involved will have approximately the same understanding of what the goal actually is.

Moreover, where an objective is vague or abstract, it is too easy to pretend that the objective has been achieved, or that progress toward it is being made, when real achievements are minimal. For example, American politicians automatically identify "freedom" with the American way of life regardless of the realities of day-to-day living in this country. Anything done to protect so-called American interests abroad is described as "defending freedom," and many Americans, probably the majority, actually accept this description.

For the foregoing reasons, it is usually true that a radical movement cannot pursue vague or abstract goals successfully. But is it *always* true? Maybe not. Look, for example, at the American Revolution. By May 1776 at the latest, the great majority of the American revolutionaries had accepted independence from Britain as their objective of highest priority.[9] This objective was clear and concrete, and it was achieved. But independence was not the revolutionaries' only goal: They also wanted to set up a "republican" government in America.[10] This was by no means a clear and concrete objective, since widely differing forms of government can be described as "republican." Consequently, once independence had been achieved, there were intense disagreements among the revolutionaries over the precise form of the "republic" that was to be established.[11] Nevertheless, the revolutionaries did succeed in setting up a government that was unquestionably republican in form and that has lasted to the present day.

Notice, however, that the revolutionaries did not set up a successful republican government until they had already won independence from Britain and no longer faced stiff opposition. Furthermore, they enjoyed certain special advantages: They had as a model a form of government—the English one—that was already halfway to being a republic. (Jefferson referred to the English constitution as a "kind of half-way house" between monarchy and "liberty."[12]) The revolutionaries shared a common heritage of relatively "advanced" political ideas derived from English tradition and from the works of Enlightenment philosophers.[13] England, moreover, had long been moving in the direction of representative democracy, so the American revolutionaries were only accelerating what was already a well established historical trend. And they were not accelerating it so very much, since the government they set up was still far from fully democratic.[14]

In Part III of this chapter we will see other examples in which movements have succeeded in reaching vague or abstract goals. But we know of no well-defined examples of this kind in which the movement has faced stiff opposition and has not been favored by a pre-existing historical trend.

It would be rash to conclude that a movement can *never* achieve vague or abstract goals against stiff opposition and without the help of a pre-existing historical trend. But it remains true that a movement that lacks a clear and concrete goal operates under a very heavy disadvantage. The stronger the opposition that a movement has to face, the more important it is that the movement should be united and able to concentrate all its energy on achieving a single objective; and this requires an objective that is clearly defined.

Yet, even in those situations in which the need for a clear and concrete objective is greatest, Postulate 1 does not imply that abstract goals are useless. Abstract goals often play an essential role in motivating and justifying a movement's concrete objective. To take a crude example, an aspiration for "freedom" may motivate and justify a movement that seeks to overthrow a dictator.

Postulate 2 is a matter of common, everyday experience. We all know how useless it is to try to change people's behavior by preaching to them—generally speaking. Actually there are some important exceptions to Postulate 2, but before we discuss those we need to point out that some seeming exceptions are not really exceptions at all.

It would be a mistake, for example, to suppose that the teachings of Jesus Christ have been effective in guiding human behavior. It seems that the earliest Christians did try to live in accord with the teachings of Jesus (as they interpreted them), but at that stage the Christians comprised only a tiny minority. With the passage of years, the Christian way of life was progressively vitiated in proportion to the growing number of Christians,[15] and by the time Christianity had become dominant in the Roman Empire few Christians still lived as those of the first century AD had done. The world went on as before, full of war, lust, greed, and treachery.

What happened, of course, was that Christian doctrines were reinterpreted to suit the convenience of the society that existed at any given stage of history. Thus, the biblical commandment barring "usury" was originally held to prohibit all lending of money at interest.[16] The prohibition was often violated, beginning at least as early as 200–250 AD, but it remained theoretically in force at least through the late Middle Ages, until it became a serious obstacle to economic development. At that point it was abandoned altogether,[17] and nowadays it would be a rare Christian who would claim that lending at interest was prohibited by his religion.

Jesus himself—if we assume that the Gospels accurately reflect his views—was opposed to *all* accumulation of wealth,[18] and the earliest

Christians probably tried to live accordingly, for "as many as were possessors of lands or houses sold them, and brought the prices of the things that were sold, and laid them down at the apostles' feet: and distribution was made unto every man according as he had need."[19] But that didn't last long. Not later than the early 2nd century AD there already were some wealthy Christians, and the Epistle of James rails against them for failing to help their impoverished brethren.[20] Over the succeeding centuries there were growing numbers of rich Christians, including many who were greedy or did nothing for the poor,[21] and today, at least in the United States, it is clear that the majority of Christians are less concerned to alleviate poverty than the (mostly non-Christian) left is.[22]

In North America and Western Europe a gentling effect—a decline in cruelty and violence—is often attributed to Christianity; Jesus is commonly seen as a pacifist. Actually Jesus's commandment, "Do not kill," was never intended to prohibit *all* killing, but only "murder," i.e., *unjustifiable* killing,[23] and Christian societies ever since have arrived at their own definitions, to suit their own needs, of what constitutes an "unjustifiable" killing, just as they would have done if Jesus had never lived. Christianity was strongest in Europe during the Middle Ages, a particularly cruel and violent era,[24] and the decline in cruelty and violence has coincided with a gradual weakening of Christianity from the 17th century to the present. So in this regard it does not appear that Jesus's teachings have had any substantial effect on human behavior.[25]

For another example, take Karl Marx. As a practical revolutionary Marx was active only for about 12 years (1848–1852, 1864–1872), and was not particularly successful;[26] his role was primarily that of a theorist, an advocate of ideas. Yet it has sometimes been said that Marx exercised a decisive influence on the history of the 20th century. In reality, the people who exercised the decisive influence were the men of action (Lenin, Trotsky, Stalin, Mao, Castro, etc.) who organized revolutions in the name of Marxism. And these men, while calling themselves Marxists, never hesitated to set Marx's theories aside when "objective" circumstances made it advisable for them to do so.[27] Moreover, the societies that resulted from their revolutions resembled the kind of society envisioned by Marx only to the extent that they were in a general way socialistic.

Marx did not invent socialism, nor did he originate the impulse to revolution. Both socialism and revolution were "in the air" in Marx's day, and they weren't in the air just because some ingenious fellow happened to dream them up. They were in the air because they were called forth by the social conditions of the time (as Marx himself would have been the first to insist[28]). If Marx had never lived there would have been revolutionaries all the same, and they would have adopted some other socialistic thinker as their patron saint. In that case the terminology and the details of the theory would have been different but the subsequent political events probably would have been much the same, because those events were determined not by Marx's theories but by some combination of "objective" conditions with the decisions of the men of action who organized the socialist revolutions. And the men of action, as we've pointed out, were guided less by Marx's theories than by the practical exigencies of revolutionary work.

Even if we assume that the political events would have been different without Marx, the events that did occur did not represent a fulfillment of Marx's ideas, because, again, the societies that grew out of the socialist revolutions did not resemble anything that Marx had foreseen or desired. So it does not appear that Marx accomplished much through his advocacy of ideas.

For similar reasons, probably very few if any of the "great thinkers" whose ideas supposedly influenced history ever achieved their goals, except where the thinkers were also men of action who were able to implement their own ideas (as in the case of the Prophet Mohammed, for example). Such thinkers, therefore, do not provide counterexamples to the principle that the advocacy of ideas, by itself, cannot produce important, lasting changes in human behavior (unless in some very small minority). Nevertheless, some exceptions to Postulate 2 should be noted.

Small children are highly receptive to the teaching of their parents and of other adults whom they respect, and principles preached to a small child may guide his behavior for the rest of his life.

Ideas that people receive may have an important, long-lasting effect on their behavior if the ideas are ones that many individuals can apply for their own personal advantage. For example, the rational methods of empirical science were at first preached only by a tiny minority, but those ideas spread and were applied throughout the world because they were of great practical utility to those who applied them. (Even so, scientific rationality is consistently applied only where it is useful to those who apply it. Scientific rationality is commonly set aside when the irrational is more useful, for example, in certain aspects of the social sciences where the goal is not to describe reality accurately but to provide support for an ideology or a worldview.)

The power-structure of a modern society can change human behavior by preaching on a vast scale through the mass media with the help of skilled professional propagandists. Maybe a group outside the established power-structure could also change human behavior through propaganda alone, but only if the group were sufficiently rich and powerful to undertake a massive, sophisticated media campaign.[29] Even where human behavior is changed by professional propagandists, however, it is doubtful that the change is ever permanent. It seems that such changes are easily reversed when the propaganda ceases or is replaced by propaganda that promotes contrary ideas. Thus, the effects of Nazi propaganda in Germany, Marxist-Leninist propaganda in the Soviet Union, and Maoist propaganda in China faded rather quickly when those systems of propaganda were discontinued.

Postulate 3. Probably every radical movement tends to some extent to attract persons who join it from motives that are only loosely related to the goals of the movement. When Earth First! was founded in the 1980s its goal was simply the defense of wilderness, but it attracted numerous individuals of leftist type who were less interested in wilderness than in activism for its own sake. A good example was the late Judi Bari, who was a radical feminist, demonstrated against U.S. involvement in Central America, and participated in the pro-choice and anti-nuclear movements. "Eventually, she added environmentalism to her list of causes"[30] and became an Earth First!er. The influx of numerous individuals of this type did lead to the blurring of Earth First!'s original mission, which became heavily contaminated with "social justice" issues.[31]

Probably, however, not every radical movement is equally attractive to persons whose goals differ from those of the movement. Because of the personal risk involved, it's not likely that an illegal and persecuted movement would draw many cranks and do-gooders, though on the other hand such a movement might be attractive to adventurers who valued danger, conspiracy, or violence for their own sake.[32] Again, when a movement is fully absorbed in a hard struggle (legal or not) for a single, specific, clearly defined goal, one imagines it would attract few individuals who were not willing to commit themselves whole-heartedly to that goal.

Whether this is true or not, it does seem true that even if many persons having varied and diffuse goals enter a movement, the movement's objective does not necessarily become blurred or perverted if that objective is simple, concrete, and clear, and if the movement is committed to it exclusively. For example, it appears that most of the early feminist leaders were professional reformers who were interested in a variety of causes, such as temperance (anti-alcohol), peace (anti-war), pacifism, abolition of slavery, and so-called "progressive" causes generally.[33] Yet, once the feminist movement had become clearly focused by about 1870 on the single, overriding goal of woman suffrage, it seems to have remained entirely faithful to that goal until the goal was achieved in the 1920s.[34]

Thus, the words "tends to" and "may" that appear in the statement of Postulate 3 signify that the postulate does not state an inviolable law, but only a danger to which social movements are subject. The danger, however, is a serious one.

Postulate 4. The meaning of Postulate 4 needs to be clarified: A movement will not necessarily be thoroughly corrupted unless it becomes so powerful that (i) membership in the movement entails little or no risk (whether of physical harm or of other negative consequences, such as drastic loss of social status); and (ii) the movement is able to offer its adherents such conventional satisfactions as money, security, positions of power, a career, or social status—meaning social status not merely within the movement but in society at large. Even then the movement's ideals may retain some residual effectiveness unless and until the movement achieves a secure position as the dominant force in society, after which corruption becomes complete.

Subject to the foregoing clarification, Postulate 4 seems to be invariably true. People who join a radical movement while it is still relatively weak may have goals that diverge from those of the movement, but at least such people are not likely to be selfish in the conventional sense, because they cannot draw the conventional advantages from their membership in the movement. In fact, their membership may entail serious risks or sacrifices. They may be motivated in part by a drive for power, but they seek to satisfy that drive through participation in a movement that they hope will become powerful and attain its goals.[35] There may also be struggles for power within the movement. But the members do not expect the safe and stable positions of power that are available in a movement that is already powerful and firmly established.

However, once a movement can offer money, security, status, a career, stable positions of personal power, and similar advantages, it becomes irresistibly attractive to opportunists.[36] At this stage the movement will already have grown to be a big one with an unwieldy administrative apparatus, so that the exclusion of opportunists will not be a practical possibility. After the Bolsheviks became masters of Russia even Lenin, powerful as he was, was unable to exclude the droves of opportunists who joined the party, and according to Trotsky these people subsequently became "one of the bulwarks of the Stalinist party regime."[37] Moreover, when a movement has grown excessively strong, even some of the formerly sincere revolutionaries may give in to the temptations of power. "The history of liberation heroes shows that when they come into office they interact with powerful groups: they can easily forget that they've been put in power by the poorest of the poor. They often lose their common touch, and turn against their own people." (Nelson Mandela)[38]

Look at history: We know very well what happened to Christianity after the Church became powerful. It seems that the corruption of the clergy has usually been in direct proportion to the power of the Church at any given time. Some of the popes have actually been depraved.[39] Islam didn't turn out any better. Twenty-four years after the Prophet's death his son-in-law, the Caliph Uthman ibn Affan, was killed by rebels, and this event was followed by power-struggles and violence among the Muslims and a prolonged period of conflict within Islam.[40] Nor does the later history of Islam indicate that it adhered to its ideals any better than Christianity did.[41] The French Revolution was followed by the dictatorship of Napoleon, the Russian Revolution by that of Stalin. After the Mexican Revolution of 1910–1920, the revolutionary ideals were progressively drained of their content until Mexico found itself under the dictatorship of a party that continued to call itself "revolutionary" without being so in reality.[42]

The sociologist Eric Hoffer wrote:

> Hitler, who had a clear vision of the whole course of a movement even while he was nursing his infant National Socialism, warned that a movement retains its vigor only so long as it can offer nothing in the present....[43]

> According to Hitler, the more 'posts and offices a movement has to hand out, the more inferior stuff it will attract, and in the end these political hangers-on overwhelm a successful party in such number that the honest fighter of former days no longer recognizes the old movement.... When this happens, the "mission" of such a movement is done for.'[44]

In March 1949, when the Communists were on the verge of final victory in China, Mao warned:

> With victory, certain moods may grow within the Party—arrogance, the airs of a self-styled hero, inertia and unwillingness to make progress, love of pleasure and distaste for continued hard living.... The comrades must be helped to remain modest, prudent, and free from arrogance and rashness in their style of work. The comrades must be helped to preserve the style of plain living and hard struggle.[45]

Needless to say, Mao's warning was futile. Already in 1957 he complained:

> A dangerous tendency has shown itself of late among many of our personnel—an unwillingness to share the joys and hardships of the masses, a concern for personal fame and gain.[46]

Today the Communist regime in China is notorious for its corruption: Not only are Party members and government officials concerned more with their own careers than they are with Communist ideals;[47] what is worse, the regime is pervaded by out-and-out criminal dishonesty.[48]

Shortly before the end of the American War of Independence, Thomas Jefferson wrote:

> It can never be too often repeated that the time for fixing every essential right on a legal basis is while our rulers are honest and ourselves united. From the conclusion of this war, we shall be going downhill.[49]

In fact, soon after the end of the war, quarreling and disunity broke out among the thirteen states to such an extent that the new nation seemed on the point of breaking up.[50] By creating the Constitution of 1787 the revolutionaries succeeded in saving the Union, but the passage in 1798 of the anti-libertarian Alien and Sedition Acts[51] suggests a weakening of commitment to the ideals of the Revolution even among some of the old revolutionaries, and by the time most of the original revolutionaries were dead not much idealism, or even integrity, seems to have been left in American politics.[52] One has to ask why the United States did not go the way of most Latin American countries and fall under the control of a dictator or an oligarchy. One part of any answer to this question should be that before the Revolution the American colonists, like their English cousins, had already been long habituated to a semi-democratic form of government, hence would not have been likely to create or tolerate a highly authoritarian regime.

III. Examination of the Rules

Because the rules are directly derived from the postulates, our discussion of the rules is in some ways merely an extension or elaboration of the foregoing discussion of the postulates.

Rule (i) asserts that a movement needs a single, clear, simple, and concrete objective.

The story of the so-called civil society movement in Mexico shows what typically happens to a movement that flagrantly violates Rule (i). The civil society movement originated in 1985,[53] and its goals were to oppose "concentrated, centralized power"[54] and to fight "for human rights, civil rights, political reform and social justice against the domination of the one-party state."[55] Thus the movement favored decentralization and "a redistribution of power,"[56] and "tended to take the side of the underdog, to side with peasants and workers, poor people and Indians."[57]

Obviously the civil society movement did not have a single, clear, concrete goal.[58] Some sectors of the movement did adopt single, clear, concrete goals. For example, the Mexican anti-nuclear movement was part of the civil society movement,[59] and its single goal was to prevent the development of nuclear energy in Mexico. It was not completely successful in achieving this, since one nuclear power-plant was put into operation in Mexico. However, "the anti-nuclear movement had really won on the question of Mexico's nuclear future," because Mexico's ruling party "abandoned its ambitious plans for a dozen or more nuclear reactors."[60]

But who hears of the Mexican civil-society movement today (2018), thirty-three years after it arose? The movement seems to have petered out without having made any significant progress toward the general goals stated above. The election in 2000 as president of Mexico of Vicente Fox of the "conservative" (read "authoritarian") PAN party may have seemed to end the "domination of the one-party state" by breaking the PRI party's monopoly of power, but many of the PRI technocrats had actually *wanted* "some sort of power-sharing arrangement with the PAN," so that Mexico would no longer appear to be a one-party state yet would remain effectively under technocratic control.[61] The technocrats' power-sharing arrangement worked up to a point: The PAN held the presidency for two six-year terms (2000–2012), after which the PRI returned to power. But on July 1, 2018 Manuel López Obrador, described as a "leftist," was elected president of Mexico, and his MORENA party won other elective offices as well.[62] Thus, the PRI-PAN system had clearly lost its grip on power. Meanwhile, however, the traditionally very powerful Mexican presidency, and with it the national government, had been to some extent weakened as increased authority was acquired by the governors of the Mexican states.[63] In this way there was a "redistribution of power" in Mexico, but it was hardly the kind of redistribution of power that the initiators of the civil society movement had had in mind: "The governors rule like 'feudal lords' with few oversights such as independent auditors and legislatures,"[64] and they are thoroughly corrupt.[65]

In another way too there has been a redistribution of power in Mexico:

> In much of the country [drug gangs are] more powerful than the government itself. Mexico's three main drug cartels are effectively in control of the country's Pacific Coast, industrial heartland, and tourist havens of the Gulf Coast. ...[T]he gangs... don't hesitate to kill the politicians, cops, and journalists they can't bribe or intimidate. ...Yet they are folk heroes to many poor Mexicans.... [The gangs'] ranks now include many members of Mexico's elite special forces. At the same time, the gangs have infiltrated much of Mexico's power structure. ...They have corrupted every level of government, from local policemen to army generals to presidential aides.[66]

But this, again, is hardly the kind of "redistribution of power" that the initiators of the civil society movement had in mind.

The new President, López Obrador, is no revolutionary. He says he wants to help poor people and no doubt he will try, but apparently he also feels it necessary to retain the support of big business,[67] and one must seriously doubt whether his efforts to help the poor will be any more effective than those of earlier Mexican presidents.[68] In any case, his pledge to "end Mexico's rampant government corruption" and put a stop to the murders of the drug gangs[69] is nothing but the kind of empty promise that politicians typically offer in order to win elections. The Mexican government has been trying for many years to bring the drug gangs under control and has made no headway;[70] "rampant corruption" is deeply ingrained in Mexico's political culture and isn't going to be eliminated any time soon.[71]

So what did the Mexican civil-society movement accomplish? Some sectors of the movement may have attained their *specific* goals, but toward the movement's *general* goals little or nothing has been achieved.[72]

In England and the United States during the first two thirds of the 19th century, the goal of feminists was to make women equal to men in terms of power, dignity, and opportunities within society. Since this goal was a vague and general one, it's not surprising that these early feminists didn't accomplish much.[73] But, as we saw earlier, by roughly 1870 feminists had settled on a single, clear, simple, and concrete objective: to secure for women the right to vote.[74] Perhaps because they realized that it was the key that would open the door to power for women and enable them to reach other goals, woman suffrage was the objective on which feminists concentrated their efforts until that objective was achieved in the 1920s.

Since the 1920s the feminist movement has had no single, clear, concrete objective. The movement has splintered into various factions that pursue diverse objectives and are often in conflict with one another.[75] Nevertheless, Rule (i) notwithstanding, feminists have continued to make steady progress toward their general goal—to make women equal to men in power, dignity, and opportunities.[76] However, the feminists have had certain critically important advantages that have offset their neglect of Rule (i).

First, the achievement of the earlier feminists' well-chosen central objective—the right to vote—has given women collective power: No politician who hopes to win an election can afford to ignore women's wants. More importantly, the tide of history has been working in the feminists' favor. Ever since the onset of the Industrial Revolution there has been a powerful trend toward "equality"—meaning the elimination of all distinctions between individuals other than those distinctions that are demanded by the needs of the technological system.[77] Thus, a mathematician is to be evaluated in terms of his/her mathematical talent, a mechanic in terms of his/her knowledge of engines, a factory manager in terms of his/her ability to run a factory, and with the passage of time it has increasingly been expected that the religion, social class, race, gender, etc., etc. of the mathematician, the mechanic, and the manager are to be treated as irrelevant. Because the feminists' goal of equality has been in harmony with this historical trend, opposition to feminism has steadily declined over time, and from 1975 at the latest the media and the cultural and political climate have been overwhelmingly favorable to gender equality.

A comparison of post-1945 British and American feminism with the Mexican civil society movement provides an illustration of the principle that the stronger the opposition a movement has to face, the more important it is that the movement should concentrate all its energy on a single, clearly defined objective. The feminists have made steady progress toward their vague goal of gender equality, in part because they have faced no very serious opposition since the middle of the 20th century. But the Mexican civil society movement has faced very tough resistance from the nation's power-structure, and the movement has therefore been doomed by its failure to concentrate on a single, clear, concrete objective.

In connection with Rule (i) it is also instructive to look at the history of Ireland. From at least 1711 until the 1880s, there was chronic rural unrest in Ireland due to the wretched conditions in which Irish peasants had to live.[78] In 1798 there was an attempt at violent revolution, but it failed miserably, in large part because it was unorganized, undisciplined, and lacked a clear objective.[79]

The Irish began to make progress only with the advent of Daniel O'Connell. O'Connell was a political genius and a spellbinding orator,[80] but unlike many other political geniuses he was a sincere patriot who had genuinely dedicated himself to the welfare of his country. O'Connell's ultimate objective was "the improvement of the lot of the Irish common people."[81] As a step toward this vague and general goal, O'Connell set himself a clear and concrete objective, namely, "Catholic Emancipation,"[82] which meant repeal of the laws that subjected Irish Catholics to certain political disabilities (for example, they were not allowed to become judges or members of Parliament).[83] Catholic Emancipation would directly benefit only a small minority who could hope to occupy important offices or be elected to Parliament, but it would indirectly benefit the overwhelmingly Catholic peasants of Ireland inasmuch as it would give them representation in Parliament and (more importantly) prove that they could prevail over the British government through collective action.[84]

O'Connell created an amazingly well-organized and well-disciplined movement dedicated to the specific goal of Catholic Emancipation, and that goal was achieved within about six years.[85] Catholic Emancipation undoubtedly would have occurred eventually anyway, since it was a development that was guaranteed by the same historical trend toward "equality" that favored the feminist movement. But, without O'Connell and his organization, Catholic Emancipation probably would have been delayed for many years, for when Emancipation was granted in 1829 it was granted grudgingly,[86] and it very likely would not have been granted at that time at all if O'Connell had not played skillfully upon the government's fear of another violent uprising like that of 1798.[87] (It is worth noting, therefore, that the 1798 rebellion, even though it was ruthlessly crushed, was not in vain.)

Needless to say, excellent organization in pursuit of a single, clear, simple, and concrete objective does not *guarantee* success. In 1840, O'Connell founded a Repeal Association for the purpose of securing the repeal of the Act of Union that placed England and Ireland under a single Parliament. The objective was not to separate Ireland from England but to create a specifically Irish Parliament, while Ireland would remain united with England under a single sovereign.[88] Again O'Connell built a highly disciplined movement that had broad support among the Irish people, but this time he failed to achieve his objective, for the British government and Parliament remained obdurate, and the Act of Union was not repealed.[89]

A contributing factor in the failure of O'Connell's Repeal Association was the Great Potato Famine of 1846–49. When peasants were starving to death in droves, O'Connell's political goal seemed irrelevant to them.[90] In 1847, during the famine, a faction within the Repeal Association formed a new organization called the Irish Confederation.[91] The new group soon recognized that it needed some specific goal,[92] but apparently was unable to agree on one until the revolutions of 1848 broke out on the European continent. Inspired by these events, the Irish Confederation adopted violent revolution as its goal, presumably for the purpose of making Ireland independent of Britain.[93] That same year an uprising attempted by the radicals failed, in part because of the radicals' incompetence, but even more because they had no popular support. The common people were concerned only with their own immediate material welfare, or indeed with their very survival, and had little interest in the Confederation's nationalism.[94]

By 1856, a leader named James Stephens (a survivor of the 1848 uprising) had definitely settled on the clear, concrete objective of total political independence for Ireland.[95] Independence was to be followed by the establishment of a "republic,"[96] but the imprecision of this second goal was perhaps not very important, because a republic would not be established until independence had been achieved. Thus, the imprecise goal of founding a republic would not necessarily interfere with efforts toward the clear and specific goal of independence. (Compare the case of the American revolutionaries, discussed above.)

Stephens, a brilliant organizer, created a powerful revolutionary movment[97] that in 1867 attempted an uprising for the purpose of separating Ireland from Britain. For reasons not relevant to the present discussion, the uprising failed ignominiously.[98] But from that time until 1916 the aspiration for total independence from Britain was kept alive by a minuscule minority of extreme nationalists who had virtually no support among the general population of Ireland.[99] Irish peasants at first were concerned only to secure relief from the oppression of the landlords, and had no interest in nationalist ideals. Eventual relief of the peasants' suffering was guaranteed by the general liberalizing trend of Western civilization, but the process was accelerated by the efforts of Parnell and Gladstone,[100] so that the condition of the peasants was alleviated step by step until by 1910 at the latest they no longer had any grievance serious enough to provide a motive for radical action.[101]

Thus, by the second decade of the 20th century, the Irish no longer had any plausible reason to separate themselves from Britain, nor did such a separation have the appearance of a historical inevitability. Nevertheless, the extremists' stubborn persistence in adhering to their goal of total independence did pay off in the end. It is a remarkable fact that between 1916 and 1921 the tiny minority of extreme nationalists, who at first lacked significant support, were able to swing the majority of the Irish population over to their side. Through terroristic tactics and guerrilla warfare, the nationalists provoked the British government to harsh countermeasures that alienated the Irish masses and drove them into the arms of the revolutionaries.[102] The result was not immediate and complete independence for Ireland. The military situation forced the revolutionaries to stop (temporarily) just short of their goal by accepting "dominion status"; that is, a relationship to Britain similar to that of Canada.[103] This made Ireland practically an independent country with ties to Britain that were little more than symbolic; and even so the revolutionaries never regarded the settlement as final, but only as a stepping stone to the total independence that was to be reached later.[104]

Nevertheless, a powerful faction of the nationalist movement, under the political leadership of Eamon de Valera, refused to accept dominion status and was suppressed only through a brief but bloody civil war.[105] A remnant of the dissident faction continued to exist, but most of it was subsequently integrated as a normal component of Ireland's parliamentary system.[106] De Valera was for many years the Prime Minister of Ireland, and by 1949 at the latest had made his country totally independent of Britain—with the ready acquiescence of the British themselves.[107]

Thus, in the end, the extreme Irish nationalists did achieve the one clear, simple, concrete goal that for many decades had been the center of their aspirations.[108] It was moreover a goal that probably would never have been reached without the nationalists' efforts, for, as noted earlier, there was no apparent historical reason why Ireland had to become independent[109] (unless the reason was the existence of the nationalists themselves).

Independence was unmistakably the extreme nationalists' dominant goal until it was very nearly achieved with dominion status in 1922. But what about the nationalists' other goals? It can probably be said of each of their other goals that either the goal would have been reached anyway through the operation of general historical trends and without any effort on the part of the nationalists; or the achievement of the goal was merely symbolic; or else the goal was achieved only in an incomplete form that would have been unsatisfactory to the original revolutionaries.

One goal of the 19th-century Irish revolutionaries was the relief of the peasants' misery, and the revolutionaries may have hastened the achievement of this goal insofar as the British fear of revolutionary violence made the task of reformers like Parnell easier.[110] But, as we've already pointed out, the eventual relief of the peasants' misery was guaranteed anyway by a pre-existing historical trend that prevailed throughout Western Europe.

Another goal of the extreme nationalists was the establishment of a "republic." This goal was a vague one since, as we mentioned in connection with the American Revolution, a wide variety of states can be called "republics." Constitutional monarchies such as Britain, Spain, or the Netherlands are not technically republics, but in practical terms their systems differ little from those of undoubted republics like France or the United States. When Ireland was officially declared a republic in 1949, little was changed;[111] "republic" was hardly more than a word or a symbol.[112] If the term "republic" were taken to mean "representative democracy," we would say that Ireland was already a republic in substance long before it became one officially. And, once the goal of independence was reached, Ireland would have become a representative democracy anyway through the operation of pre-existing historical trends, just as every other country in Western Europe has become a representative democracy. And it is not certain that Ireland has become a republic in a sense that would have satisfied the original revolutionaries, for at least some of these seem to have had something more socialistic in mind.[113]

In addition, the revolutionaries wanted to avoid "Anglicization" of Ireland and preserve Irish language and culture.[114] In this it can't be said that the revolutionaries failed completely, but their success has been unimpressive at best. Irish Gaelic today is the first language of only a small fraction of the Irish population. Though it is taught in the schools and is "more widely read, spoken, and understood [as of 2003] than during most of the 20th century,"[115] it seems unlikely that the majority of modern Irish people are fluent in Gaelic; for, among people who learn a language only in school, no more than a small minority ever become fluent in it. Ireland is basically an English-speaking country, from which it follows that Ireland must be subject to considerable cultural influence from other English-speaking countries. Whether or not it is accurate to speak of an "Anglicization" of Ireland, there can be no doubt that Ireland (with the possible exception of a few isolated areas that may not yet be fully modernized) has undergone the same cultural homogenization that has occurred in the rest of Western Europe, and in all probability modern Ireland differs culturally from other Western European countries no more than these countries differ among themselves. It may well be that traditional arts, crafts, music, etc. are now practiced in Ireland more than they would have been without the efforts of the nationalists, but it is certain that the basic culture of Ireland today is the universal culture of modern industrial society.[116] It follows that traditional arts and crafts can be no more than gimmicks that serve to entertain tourists or to give the Irish themselves a temporary illusion of escape from the modern world.[117]

Would this degree of linguistic and cultural preservation have satisfied the revolutionaries of the 19th and early 20th centuries? Probably not. "Few Irishmen today would accept that what Irish nationalists have achieved represents a true fulfillment of that near-mystical ideal for which, in one form or another, Irishmen had striven for so long."[118]

Thus, while it can't be said that the Irish revolutionaries achieved no success at all in other areas, their one unmistakable and complete success was in reaching the single, clear, simple, and concrete goal that had been their main objective for several decades: to make Ireland politically independent of Britain.

The examples of feminism and Irish nationalism (among others) show that Rule (i) cannot correctly be understood to mean that no social movement can ever achieve any success at all without concentrating exclusively on a single, clear, simple, concrete objective. But these and other examples that we've looked at do support the proposition that every movement that hopes to achieve something had better give careful consideration to Rule (i) and should deviate from the rule only if there is a definite, strong, and convincing reason for doing so.

Rule (ii) states that if a movement aspires to transform a society, then the objective selected by the movement must be of such a nature that the social changes wrought by the achievement of the objective will be irreversible—meaning that the changes will survive even without any further effort on the part of the movement or anyone else. The reason is that the movement, once in power, will become "corrupt"; i.e., will no longer be faithful to its earlier goals and ideals.

For example, the feminists' achievement of woman suffrage is irreversible because (among other reasons) now that women have the right to vote, it cannot be taken away from them through democratic processes without the consent of most women—consent that they would hardly be likely to give even in the absence of an effective and uncorrupted feminist movement. Of course, the feminists' achievement is not irreversible in any absolute sense. Women could lose the right to vote in the event of some sweeping transformation of society, such as an end of the democratic form of government.

The theocratic republic set up in Geneva by Calvin[119] provides a probable example in which a movement achieved its objective and the associated social changes were later reversed due to corruption of the movement. What seems to happen more often, though, is that a movement becomes corrupt *before* it reaches its objective, so that the objective is never fully achieved in the first place. Thus, in Russia, the Bolshevik/Communist movement was corrupted while the construction of a socialist society was still in its early stages, so that the kind of socialist society envisioned by the original Bolsheviks was never attained.[120] The French Revolution was corrupted before it even came close to creating the type of society envisioned by any of the revolutionary factions.

This writer is not aware of even one unarguable example in which a revolutionary movement has concentrated its efforts on a *single, clear, simple, concrete* objective (as required by Rule (i)) and has achieved the objective, and the achievement has subsequently been reversed due to corruption of the movement. Once a clear, simple, concrete objective has been achieved, its achievement no doubt is less easily reversed than that of a vague or complex objective, because its reversal would be too obvious, too hard to disguise. This is another reason why a movement should obey Rule (i).

The establishment of a democratic government is not a very clear and precise objective, because there are considerable differences among the various kinds of government that today are called "democratic." But democratic government is at least a much clearer objective than such vague goals as "freedom," "equality," "justice," "socialism," or "protecting the environment." From the history of many countries scattered around the world we know how reversible the achievement of democracy can be. The overthrow of a democratic government through a military coup was once such a common event in Latin America and Africa that news of such a coup hardly raised an eyebrow in Western Europe or the United States. A military coup usually represents not the corruption of democracy but a victory of those who never wanted democracy in the first place. But the death of a democracy through corruption (in our sense of the word) has probably been even more common than the military coup, and when this happens the external forms of democracy often are retained even while an individual or an oligarchy takes effective control of the country. We've seen this in Russia since the breakup of the Soviet Union: Vladimir Putin was originally a protégé of Boris Yeltsin, the great champion of Russian democracy, and Russia still retains all the usual apparatus of parliamentary democracy. Yet it is said that Putin is now almost a dictator.[121]

In Latin America, democracy has routinely been corrupted. A group of Argentine scholars presents the following example as typical:

> The center of gravity of control passed softly and silently from the State to a restricted economic and social apparatus (inside groups) that constituted a privileged system of enrichment. The acquisition of a license for importation or for exchange made more men rich in less time than any other activity, including speculation in land. This is how the families dominant in the financial sector entered political channels.... There occurred a metamorphosis of the 'group of families,' which, incidentally allied for the purpose of negotiating with the State... turned into a stable, indissolubly united oligarchy for control of the organisms of political activity... in the parties, in the government... and in order to establish a *system* of privilege.[122]

A democratic government can easily be subverted because it is a complicated mechanism and the concept of "democracy" itself is far from precise, so that subtle or covert changes can accumulate over time until one day people wake up to find that their country is no longer a functioning democracy. Compare democracy with the clear and simple achievements of the feminists and the Irish nationalists—the right to vote and political independence for Ireland, respectively. Because of the clarity and simplicity of these achievements, they could not easily be undermined covertly. Note, however, that while impairment of Ireland's *formal political* independence would be obvious, Ireland could become dependent on Britain economically or in some other way.

Though the establishment of democratic government has so often been a reversible social change, it cannot be said that democratization per se is reversible. Whether it is reversible or irreversible in a given case depends on the culture and history of the country in which the democratic government is set up, and on the international situation in which the country finds itself. The more or less democratic system set up by the American revolutionaries survived despite the fading of revolutionary idealism, in part because the American colonists had already been long habituated to a semi-democratic form of government. Today in Latin America functioning democratic governments seem to have a better chance of success than they did a few decades ago, probably because the cultural and economic changes associated with modernization have raised the level of social discipline in those countries.[123] Another factor to be considered is that the international climate has become more unfavorable to obvious dictatorships (by an individual or by a party), and nations are now under pressure to maintain at least the appearance of democracy. In Africa, for example, international aid organizations left The Gambia after a military coup in 1994 but resumed assistance to that country following its return to democratic forms;[124] assistance to Tanzania from the International Monetary Fund apparently was conditioned on political reforms in 1986;[125] in Kenya, "Western financial aid came to be tied to demands for political and economic reforms," so that in 1991 there was a "constitutional amendment that reinstated multiparty elections."[126] On the other hand, it must seriously be doubted whether democracy is a functioning reality in those countries. Kenya,[127] at least, is not a democracy in the sense in which that term is understood in Western Europe and the United States. The Gambia—notwithstanding democratic forms—was ruled by a dictator from 1994 to 2017.[128] It appears that the country is now (2018) making another attempt at democracy—with what success remains to be seen.

As the foregoing discussion indicates, the problem of predicting whether a social change will be reversible or irreversible in a given case can be subtle and difficult. Consequently, Rule (ii) may often be hard to apply in practice. Rule (ii) is important nevertheless. The essential point of the rule is that a movement builds its foundation on quicksand if it bases its strategy on the assumption that faithfulness to the movement's ideals will be sustained indefinitely and independently of the immediate self-interest of the people who are in positions of power. What a nascent movement needs to ask itself in choosing its objective is whether the resulting social changes, once achieved, will survive in an atmosphere in which people are motivated more by short-term self-interest than by dedication to ideals—which indeed is the normal atmosphere in any society. Even though this question may often be difficult to answer with any degree of confidence, it needs to be asked and considered carefully.

Rule (iii) states that once an objective has been selected, some small minority must undertake organization for practical action (as opposed to mere preaching or advocacy of ideas) in the service of the objective.[129] Three points must be noted, however:

First, while ideas *by themselves* will not transform a society, the development and propagation of ideas must be a *part* of any rational effort to transform a society. Without some organized set of ideas to guide its action, a movement will flounder aimlessly. It may generate more or less uproar, but if it accomplishes anything more than that it will do so merely through luck.

The Whiteboy Movement of 18th-century Ireland consisted of guerrilla-like peasant bands that roamed the countryside at night taking revenge on the landlords and on those peasants who were too ready to cooperate with the landlords.[130] But the members of these bands were uneducated men whose limited ideas did not enable them to envision anything beyond resistance to specific local abuses.[131] Only in the 1790s, with the arrival of ideas from revolutionary France, did the Irish peasant rebels begin to acquire some notion of changing society.[132] At the time of the attempted revolution of 1798 their ideas in this direction were still too confused to provide them with a clear objective,[133] and their lack of a clear objective was probably a contributing factor in their defeat.[134]

Contrast the Irish peasants of 1798 with the workers of Saint Petersburg who revolted in February 1917: These workers had already been indoctrinated with Marxist ideas by the Bolsheviks, consequently their insurrection was purposeful and successful.[135]

Second, while both ideas and organization for practical action are necessary components of any rational and successful effort to change a society, the people who organize for practical action need not be the same individuals as the theorists who develop and propagate the ideas. In Ireland, again, nationalist ideas and the aspiration for independence from Britain were already well developed among the extremist minority prior to the advent of Michael Collins in 1917.[136] Collins does not seem to have been a theorist, but it was he who organized the successful guerrilla war that led to Ireland's independence in 1922.[137]

However, for theorists who do not themselves organize for practical action, there is a grave danger: The men of action who do organize, purportedly in the service of the theorists' ideas, may reinterpret or distort the ideas so that the results are very different from what the theorists envisioned. Martin Luther was appalled at the social rebellion that his ideas called forth,[138] and we've already pointed out that Marxist revolutionaries like Lenin, Trotsky, Stalin, Mao, and Castro deviated from Marx's ideas whenever they found it convenient to do so. Here again we see the importance of Rule (i), that is, of the need to select a clear, simple, concrete objective: Neither Marx nor Luther formulated such an objective, and because their ideas were complex their ideas could easily be misunderstood or distorted. In contrast, by the time Michael Collins assumed leadership of the Irish nationalist movement, the nationalists had already settled on total political independence from Britain as their central objective—an objective of such clarity and simplicity that it could hardly be misunderstood or distorted.

Third, preaching, or the advocacy of ideas, is by far the easier part of an effort to change a society; organizing for practical action is vastly more difficult. This at least is true today; it may not always have been true in the past.

Martin Luther was an intellectual leader, but he was not a man of action; he even declined to try to carry out the "institutional church reforms" that he himself had called for.[139] Yet his preaching and his daring theological ideas aroused tremendous ferment, as a result of which armies were organized and wars fought.[140] It appears that organization for practical action occurred quickly and easily once Luther's ideas became widely known. In those days the educated sector of society was relatively tiny, and the expression of dissident ideas could entail considerable personal risk.

(Luther's forerunner Jan Hus was burned at the stake for his ideas.[141]) Consequently, new ideas were a scarce commodity, and intense dissatisfactions could long fester unarticulated for lack of anyone to articulate them. A thinker who was bold enough to express dissent publicly and able to do so eloquently might trigger a release of pent-up resentments. When this occurred it was probably much easier to organize a rebellion in those days than it is now, because people were much less effectively conditioned to obedience, docility, and passivity than they are today. In fact, by modern standards the people of Luther's time were lawless.[142]

Nowadays, however, there is a surfeit of ideas, including dissident and even outrageous ones. Artists and writers strive to outdo one another in thumbing their noses at conventional values. Consequently new ideas, however outrageous, evoke a yawn from many people, from others only an expression of irritation, and serve the remainder of the population as mere entertainment. To their contemporaries, the ideas of men like Hus and Luther suggested the possible opening of a new era, but no ideas do that today because new ideas are so commonplace that no one takes them seriously any more. Except, of course, technological ideas.

At present, organization for practical action is more difficult not only because new ideas no longer evoke a strong response, but also because of people's docility, passivity, and "learned helplessness."[143] Professional political operatives do exploit people's discontents to organize support for their parties, candidates, or movements, but this only makes the task of organization more difficult for amateurs, who are poorly equipped to compete with skilled professionals for people's attention and commitment.

Thus, whatever may have been the case in the past, in the modern world the critical challenge for anyone wishing to transform society is not the propagation of ideas, but organization for practical action.

Rule (iv) states that in order to keep itself faithful to its objective, a movement should devise means of excluding all unsuitable persons who may seek to join it.

One can identify two possible approaches to the problem of excluding unsuitable persons: (a) A movement may be careful in selecting the individuals who are allowed to be members; or (b) a movement may design its program or its public message in such a way that unsuitable persons will not *want* to be members. As far as this writer's knowledge extends, relevant historical evidence is scanty. Only in the case of communist movements does there appear to be any indication of explicit policies intended to exclude unsuitable persons; though no doubt many movements have to some extent deterred any influx of unsuitable persons in one or both of the foregoing ways without adopting explicit policies for that purpose. Unsuitable persons may simply have been given the cold shoulder at meetings, or the movement's program and message, even without any conscious intention, may have tended to repel unsuitable persons.

The Bolsheviks seem to have used both of the approaches (a) and (b) to the problem of excluding unsuitable persons. Lenin "was always extremely sensitive to the question of the ingredients of the party;"[144] he insisted on "strict selection of members" and that "an organization of real revolutionaries will stop at nothing to rid itself of an undesirable member."[145] This writer has no information as to the means by which "strict selection of members" was to be carried out or how "undesirable members" were to be identified and ejected from the party. In any case, Lenin also insisted that every member of the party should be required to belong to a party organization and submit to party discipline,[146] in part because, under this requirement, "unstable elements" would not *want* to be members of the party.[147]

In the years preceding the Communist victory in China, Mao repeatedly referred to the problem of weeding out the "careerists," "saboteurs," "degenerates," "undesirables," and "traitors" who "sneaked" into the Party.[148] But nowhere in the *Selected Readings* from his works does Mao explain what he means by "saboteurs," "degenerates," etc., nor does he tell us how these individuals are to be identified and excluded.

Beyond the foregoing, this writer knows only that the Bolsheviks and the Chinese Communists did avoid letting any unsuitable elements in their organizations become numerous enough to prevent these two movements from seizing power in their respective countries. *How* they avoided letting unsuitable elements become numerous remains obscure, though they were no doubt helped by the fact that—as we pointed out earlier—an illegal and persecuted movement is unlikely to attract many cranks and do-gooders.

In 1841 there was an attempt to set up a utopian community at Brook Farm in Massachusetts. Though the community was linked with the distinguished intellectuals of the Transcendental Club it failed within a few years,[149] perhaps in part because it was joined by too many of "the conceited, the crotchety, the selfish, the headstrong, the pugnacious, the unappreciated, the played-out, the idle, the good-for-nothing generally;

who, finding themselves utterly out of place and at a discount in the world as it is, rashly concluded that they are exactly fitted for the world as it ought to be."[150] It has been suggested that the Brook Farm experiment would have had a better chance of success if "standards of recruitment" had been applied.[151] No doubt standards of recruitment—if there had been any—could readily have been applied in this case because Brook Farm was a relatively small enterprise at a single location, so that a handful of leaders would have been able to evaluate any prospective member directly by interacting with him personally. But once an organization has grown larger and has, say, several thousand members at scattered locations, the task of vetting prospective recruits will have to be delegated to various lower-level leaders who necessarily will exercise a good deal of independent judgment. Under these circumstances, it may be very difficult to ensure consistent application of rigorous standards of recruitment.

During the Stalinist era, communist parties in non-communist countries were selective in accepting members,[152] and they seem to have succeeded in maintaining adequate standards in doing so. But these parties were tools or agencies of the Soviet Union, centrally governed from Moscow,[153] and they had a well-established, tightly disciplined, hierarchical structure.[154] A new and growing movement, independent of any strong central authority, will not easily build a tightly disciplined structure and consequently will find it difficult to maintain consistent standards in selecting individual recruits. We therefore suggest that such a movement should emphasize the approach (b) mentioned above—it should make a special effort to design its program and its public message in such a way as to repel unsuitable persons who might otherwise seek to become members.

A movement that follows Rule (i) will perhaps have taken a step toward so designing its program; as we noted in the discussion of Postulate 3, it seems unlikely that a movement struggling toward a single, specific, clearly-defined goal will attract many individuals who are not willing to commit themselves whole-heartedly to that goal.

The fact that the 19th-century feminist movement lent its support to Victoria Woodhull, who was a spiritualist charlatan and an advocate of a crackpot variety of socialism,[155] suggests that that movement may not have been selective in accepting individual participants. If that is true, then the movement may have been saved by its focus on the single, specific, clearly-defined goal of woman suffrage.

In relation to Rule (iv) there is also the question of where to draw the boundary of the "movement" from which unsuitable persons are to be excluded. If a movement has an inner circle and an outer circle, and if the inner circle maintains firm control over the outer circle, then exclusion of unsuitable persons from the inner circle may be sufficient even if all comers are admitted to the outer circle.

It's very probable that the Irish patriot Daniel O'Connell had his movement for Catholic Emancipation firmly under his own personal control for the six years of its existence,[156] 1823–29. But whatever may have been the degree of O'Connell's personal control over it, his superbly organized and highly disciplined[157] movement must have been governed by some limited inner circle that presumably was kept "pure" through formal or informal selectiveness in recruitment. Recruitment to the outer circle of the movement was indiscriminate—anyone could join by paying minimal dues[158]—but the controlling inner circle evidently kept the movement faithful to its objective.

Referring to the period about 1908–1912, the historian writes of the extreme Irish nationalist movement:

> Sinn Fein...[was] a rallying point for all radical, dissatisfied and potentially disappointed individual nationalists in Ireland. The hybrid nature of its support with an overlap of poets, eccentrics, members of the Irish Republican Brotherhood, politically-minded Gaelic Leaguers and frustrated parliamentarians had been the movement's chief characteristic. Many lone wolves with a romantic or otherwise obsessional love of Ireland that had been born of Irish history, but frustrated in the present, gravitated towards Sinn Fein. As with every movement that attracts rebels there were those who [were impelled by] obscure psychological motives of their own.[159]

Evidently, then, any Irishman "opposed to British rule in Ireland"[160] could participate in Sinn Fein, and like many other radical movements it attracted a motley assortment of oddballs. But it is possible to identify with a reasonable degree of confidence the factors that rescued the movement from impotence. First, the movement was focused on the single, clear, simple, concrete objective of total political independence for Ireland,[161] and, as noted earlier, such an objective is not as easily perverted as a more diffuse one. Moreover, beginning in 1917, Michael Collins with a limited inner circle of collaborators progressively took over effective control of the movement,[162] and the inner circle, because it was relatively small, was probably easy to keep "pure" (i.e., faithful to the movement's objective) even without formal standards of recruitment. It's true that Collins and his inner circle were by no means able to control in detail the actions of their guerrilla fighters,[163] but involvement in a guerrilla war was itself a powerful factor in keeping the movement faithful to its objective. If a movement is locked in a desperate struggle, that fact will tend strongly to unite the movement behind its leaders and behind its principal objective.[164]

In summary, this writer has found very little evidence concerning any formal or informal, conscious or unconscious means that may have been used to exclude unsuitable persons from radical movements of the past. It is clear, however, that the kinds of people who join a movement necessarily have a profound effect on its character and can blur or change its goals. If some of the movements we've looked at have remained faithful to their goals without any premeditated effort to exclude unsuitable persons, then they've been lucky. A nascent movement that is not content to depend on luck needs to give close attention to the question of the kinds of people who are to comprise the movement.

Rule (v) states that once a revolutionary movement has become powerful enough to achieve its objective it must achieve its objective soon thereafter, before the movement is corrupted (as Postulate 4 affirms it will be).

As noted in the discussion of Postulate 4, this writer has found no exception to the law that when a radical movement grows too powerful it is soon corrupted; that is, it ceases to be faithful to its original goals and ideals. From this law the importance of Rule (v) is obvious. It will nevertheless be instructive to see how Rule (v) relates to some of the examples we've looked at.

In Russia, the kind of socialism envisioned by the revolutionaries could not have been built within any brief period. Consequently, as pointed out in the discussion of Rule (ii), the construction of a socialist society was still in its early stages when the Bolshevik/Communist movement was corrupted, with the result that socialism as conceived by the original Bolsheviks was never achieved at all.

It seems that democratization movements in any country, once they've achieved power, usually set up representative democracies soon thereafter. (Whether these democracies survive is another question, as we've seen.) But the French revolutionaries of the 1790s were unable to set up a properly functioning democratic government promptly. The exact point at which the French Revolution was corrupted may be open to argument, but certainly it had been corrupted by the time Napoleon became First Consul. When that happened, it was too late to establish a representative democracy.

In Mexico following the revolution of 1910–1920, the revolutionaries did not bring social justice to the peasants all at once but sought "a more conservative evolution... and more stability in government."[165] Progress toward social justice for the peasants essentially ended in 1940 when Lázaro Cárdenas, one of the original revolutionaries, concluded his term as president. Thus, delay in fulfillment of the revolutionary ideal prevented its complete fulfillment. Even the partial fulfillment of the ideal that had been achieved was eroded under attacks that began almost immediately after Cárdenas left office, and continued until the coup de grace was administered by President Salinas de Gortari (1988–1994).[166]

In England and the United States the feminist movement achieved its central goal—woman suffrage—as soon as it was powerful enough to do so. Since then the feminist movement, though splintered into various factions, has remained sufficiently powerful to make continued progress toward total equality for women, as described earlier in this chapter, but, as far as this writer knows, the movement has not been seriously corrupted in the sense of allowing the personal ambitions of its members or its leaders to supersede the movement's ideal of equality of the sexes.

However, Rule (v) refers to *revolutionary* movements, and feminism today is not a revolutionary movement. When it emerged during the first half of the 19th century feminism might perhaps have been called revolutionary, since immediate implementation of the feminists' demands would have entailed a fairly radical alteration of society. But, as noted earlier, feminism was favored by the historical trend toward "equality" in general, and by the time feminists acquired the right to vote in the 1920s their movement could no longer be considered revolutionary; one would hardly say that the achievement of woman suffrage caused a social earthquake. Still less is the feminists' goal of total gender equality a revolutionary one nowadays.

Because its goals have not been of revolutionary magnitude, the feminist movement has not had to grow powerful enough to become attractive to opportunists. Membership in feminist organizations today does not in any substantial degree earn a woman such personal advantages as money, power, or social status.[167] A woman seeking such advantages will enter a career in business, government, politics, or the professions, not in a feminist organization. Thus Postulate 4 and Rule (v) do not apply.

In the case of the Irish nationalist movement the issue is somewhat complicated. By achieving dominion status in 1922 the nationalists reached the main part of their goal, and they reached it as soon as they were powerful enough to do so.

It's not clear that it would be reasonable to say that the movement thereafter became corrupt in the sense of being no longer faithful to the goal, because the goal was already mostly achieved. Yet, once the movement was in power, it split into two factions over the oath of allegiance to the British crown that members of the new Irish Parliament had to take.[168] The more extreme faction, led by Eamon de Valera, regarded the members of the other faction (which accepted the oath) as sell-outs (as "corrupted" in our sense of the word) for giving in to the British on this largely symbolic issue.[169]

De Valera's faction eventually came into power, but nevertheless remained faithful to its goal of total independence from Britain until, by 1949, the objectionable oath had been eliminated, Ireland had been formally declared a republic, and the last vestiges of political dependence on Britain had disappeared. But all this took place under the leadership of de Valera,[170] who was one of the original revolutionaries.[171] Postulate 4 does not assert that a successful revolutionary movement is corrupted until all of its original leaders have become politically inactive.

Moreover, in another sense it could be argued that even de Valera's faction of the Irish nationalist movement was corrupted, for a certain fraction of Ireland ("Northern Ireland") remains tied to Britain even today as part of the United Kingdom.[172] The original Irish revolutionaries regarded such a partition of their country as unacceptable; their goal was independence for *all* of Ireland.[173] At least until 1998, the Republic of Ireland maintained a nominal claim to sovereignty over Northern Ireland, but there were no efforts on the part of mainstream Irish politicians to make that claim effective.[174] These politicians, like politicians everywhere, have no doubt been concerned primarily with their own careers. (They are "corrupt" in our sense.)

Thus, having been unable to take Northern Ireland from the British soon after they acquired power, the Irish nationalists lost that territory forever, or at least for the foreseeable future.[175] There still exist offshoots of the original Irish nationalist movement[176] (Sinn Fein and the IRA, the Provisional IRA, the Real IRA, the Effective IRA, or whatever the latest faction of a faction of a faction is called) that may be uncorrupted in the sense of remaining faithful to the goal of independence for *all* of Ireland, but these offshoots do not have great power, hence Postulate 4 does not apply to them.

The Reformation was not the work of a single movement but a complex event in which several theological movements, such as those of Luther, Zwingli, and Calvin, competed with one another,[177] and in which various princes participated for reasons that likely had more to do with their own practical advantage than with religious conviction.[178] Thus an examination of the Reformation in relation to Postulate 4 and Rule (v) would be complicated and would require a detailed knowledge of the period.

* * *

As we've seen from the examples reviewed here, our five rules are not to be taken as rigid laws that every radical movement must consciously obey on pain of total failure. In many situations the interpretation of the rules may be difficult and complicated, or the application of some of the rules may be impossible or unnecessary. The rules nevertheless are important because, at the least, they set forth problems that every radical movement needs to study carefully. A movement that does not consciously address the problems represented by the rules may possibly succeed through mere luck, but its chances of success will be very much less than those of a movement that takes the rules into consideration.

In the next section we will see how present-day efforts to deal with the problems generated by modern technology, including the problem of environmental devastation, are doomed to failure through neglect of the five rules.

IV. The Application

Let's start with Chellis Glendinning's "Notes Toward a Neo-Luddite Manifesto," which can be found in an anthology compiled by David Skrbina.[179] Glendinning's statement of the goals of neo-luddism is long and complicated, and most of the stated goals are hopelessly vague. Here is a sample:

We favor the creation of technologies in which politics, morality, ecology, and technics are merged for the benefit of life on Earth:

- *Community-based energy sources* utilizing solar, wind, and water technologies—which are renewable and enhance both community relations and respect for nature;
- *Organic, biological technologies*... which derive directly from natural models and systems;
- *Conflict resolution technologies*—which emphasize cooperation, understanding, and continuity of relationship; and
- *Decentralized social technologies*—which encourage participation, responsibility, and empowerment.

...*We favor the development of a life-enhancing worldview in Western technological societies.* We hope to instill a perception of life, death, and human potential into technological societies that will integrate the human need for creative expression, spiritual experience and community with the capacity for rational thought and functionality. We perceive the human role not as the dominator of other species and planetary biology, but as integrated into the natural world with appreciation for the sacredness of all life.

One can hardly imagine a more flagrant violation of Rule (i), which states that a movement needs a single, clear, simple, concrete goal. Nor is this a case in which vague, generalized goals may be attainable because a movement faces no serious opposition and is favored by a pre-existing historical trend. On the contrary, modern society is driven hard along its present technological path by the vigorous, determined, unremitting efforts of innumerable, deeply-committed scientists, engineers, and administrators, and by desperate competition for power among large organizations. Under these circumstances, the vagueness and complexity of Glendinning's goals are by themselves sufficient to guarantee the failure of her proposals.

What about Rule (v), which requires that a successful revolutionary movement achieve its goals promptly, before corruption sets in? As the basis for a thorough reorganization of society (radical enough to be called a revolution even if nonviolent), Glendinning's proposal demands the creation of a broad range of technologies, most of which differ widely from any well-developed technologies that exist today. The creation of these technologies, if possible at all, would require extensive, systematic research, vast resources, and a great deal of time. A neo-luddite movement would be able to gain control over the resources it needed only if it became big, powerful, and well-organized, hence ripe for corruption. In order to carry out the necessary social reorganization, the movement would even have to be the dominant force in society, and the process of reorganization would surely take at least a few decades—say forty years at a minimum. By that time the movement's original leaders would all be out of action and the movement would be corrupt, as guaranteed by Postulate 4. Consequently, the reorganization of society in accord with neo-luddite principles would never be completed.

Let's nevertheless make the improbable assumption that society had been transformed in the way advocated by Glendinning. Would the transformation be irreversible, as Rule (ii) requires? That is, would society remain in its transformed condition without continuing effort by the neo-luddites? Not a chance! As discussed in Chapter Two, natural selection guarantees that conflict and competition for power would re-emerge after the neo-luddite utopia had been established. Even if one rejects the argument of Chapter Two, it is an observable fact that human affairs have usually if not always been characterized by conflict and competition, whether within societies or between different societies. Glendinning does not explain what would prevent conflict and competition from reappearing and wrecking the neo-luddite utopia. In practice, the neo-luddite movement would be corrupted, just as every other radical movement that has become the dominant force in a society has been corrupted. Neo-luddite ideals would be forgotten or would receive only lip-service, and the continued existence of modern technology (which Glendinning does not contemplate eliminating) would ensure society's inevitable return to its present destructive trajectory.

As for Rule (iii), Glendinning shows no awareness of the need to form an organized movement committed to practical action. Apparently, either she thinks she and other neo-luddites can transform society just by preaching, or else she hopes someone else will do the hard work of organizing an effective movement. As we noticed earlier, the advocacy of ideas is easy; what is difficult is the task of organizing for practical action. Confronted with this task, people like Glendinning feel intimidated. They are appalled at the catastrophic growth of the technological system and they want to do something about it, but they are too helpless and ineffectual to face up to the formidable challenge of building a movement. So to give themselves the illusion that they are "doing something" they preach about the way they think we should deal with technology or with the devastation of our environment. The result is that we have an abundance of dreamy utopian schemes for saving the world, but in practical terms nothing gets done.

There are of course groups that do organize themselves in pursuit of fairly definite goals of limited scope; for example, groups like the Sierra Club that try to preserve wilderness. And they do accomplish something—a little bit—but what they accomplish is insignificant in relation to the problem of technology in general. The insignificance of their accomplishments is guaranteed by the limited scope of their goals.

Since Glendinning doesn't even mention the need to form an organized movement, the question of Rule (iv) (that a movement should find means of excluding unsuitable persons) does not arise.

But the worst is that Glendinning is utterly naïve; she doesn't even show any awareness that the problems indicated by Rules (i) through (v) exist. Her neo-luddite scheme therefore is no better than any of the other unreal utopian fantasies that have misled the unwary ever since Plato dreamed up his ideal republic.

Skrbina's anthology also contains an essay by Arne Naess, the Norwegian philosopher who coined the term "deep ecology."[180] Taken simply as criticisms of the technological system, many of Naess's remarks are quite valid. But it appears that Naess wants to bring about far-reaching, fundamental changes in the way the system functions in the real world, and to the extent his ideas are intended to lead us toward that practical objective, they are totally useless.

Naess's goals are—if such a thing is possible—even more diffuse than those proposed by Glendinning. In fact, Naess in this essay does not explicitly enumerate his goals at all. But he does write:

> A crucial objective of the coming years is... decentralisation and differentiation as a means to increased local autonomy and, ultimately, as a means to unfolding the rich potentialities of the human person.[181]

The ultimate goal, "unfolding the rich potentialities of the human person," is just beautiful; one can hardly conceive of a more elegant platitude. But as a practical proposal it is meaningless. The intermediate goals of "decentralisation" and "local autonomy" are not meaningless, but they are still too vague to form the basis for an effective movement.

Naess also writes that it is "a major concern to find a kind of equilibrium" between "the requirements of reduced interference with nature and satisfaction of human vital needs."[182] This does not even remotely approach the degree of specificity that a goal must have in order to be practical. Naess does slightly better when he quotes eight pairs of related goals stated by Johan Galtung.[183] Two of the pairs are:

> Clothes [:] build down international textile business [—] try to restore patterns of local handicraft: symbiosis with food production
>
> Transportation/communication [:] less centralised, two-way patterns, collective means of transport [—] try to restore patterns of walking, talking, bicycling, more car-free areas, cable TV, local media

Most of Galtung's goals are still too vague to serve as the basis for an effective movement, but some at least are definite enough so that individually they might serve as starting points from which one could try to develop more precise goals. However, eight pairs of goals are too many; and the achievement even of every one of Galtung's goals would not be anywhere near enough to solve the overall problem of technology. Thus, Naess's scheme violates Rule (i) as flagrantly as Glendinning's does.

Naess is ignorant of Rule (v): He thinks "big, centralised, hierarchical" social structures can be "phased out gradually."[184] Evidently he envisions a transformation of society that is to take at least a couple of generations; but in that case "deep ecology" will be corrupted long before the transformation is complete. Once "deep ecology" has been corrupted, people in positions of power will pursue primarily their own advantage and will use "deep ecology" concepts only as propaganda if they use them at all. So the transformation envisioned by Naess will never be completed.

Naess's scheme also violates Rule (ii): Even if society had somehow been transformed in the way Naess desires, the transformation would not be irreversible. It seems clear that Naess expects the retention of a good deal of advanced technology,[185] and constant vigilance would be necessary to prevent that technology from being used in ways that were inconsistent with the kind of society that Naess proposes. In practice, such vigilance would not be long maintained, because corruption (in our sense of the word) inevitably would set in.

As for Rule (iii), Naess, like Glendinning, seems to think he can save the world just by preaching, for he gives no indication of any awareness of the need to organize the "deep ecology" movement for practical action.

* * *

We could review the work of other writers in this genre—Ivan Ilich, Jerry Mander, Kirkpatrick Sale, Daniel Quinn, John Zerzan, the whole useless crew—but there would be little point in doing so, because we would only be repeating the same criticisms that we've directed at Glendinning and Naess.[186] This entire body of literature suffers, by and large, from the same faults as the work of these last two writers: Authors express their well-grounded horror at what the technological system is doing, but the remedies they suggest are totally unrealistic. There are many reasons why their remedies are unrealistic; in the present chapter we've discussed only those reasons related to the dynamics of social movements as reflected in our five rules, but in Chapters One and Two, and elsewhere,[187] we've described other very powerful reasons why solutions like those of Glendinning, Naess, Illich, Mander et al can never be put into practice.

The reader may well ask whether it is possible to conceive of any remedy at all for the problem of technology that would be consistent with the five rules. We think it is possible. To begin, let's follow Mao's advice and ask what is the principal contradiction in the situation with which we are faced. The principal contradiction, clearly, is that between wild nature and the technological system. This suggests that the objective chosen should be that of "killing" the technological system as we've described previously.[188] In other words, revolutionaries should aim to bring about the collapse of the system by any means necessary.

Rule (i): This objective is sufficiently clear, concrete, and simple to form the basis for an effective movement.

Rule (v): If a revolutionary movement once grew powerful enough to destroy the technological system in this way, it ought to be able to accomplish the destruction in a short time. Destruction is easier by far than construction.

Rule (ii): If the system were thoroughly broken down the effect would be—at least for a long time—irreversible, because it would take several hundred years or more for a new technological system to develop.[189] Some people even believe that a technological system could never again be created on Earth.[190]

Rule (iv): A revolutionary movement aspiring to "kill" the technological system would need to find a way of preventing unsuitable persons from joining the movement. Most likely the chief danger would come from people of leftist type (as defined in ISAIF[191]) who attach themselves to "causes" indiscriminately.[192] A movement could probably drive such people away by maintaining a continuous verbal and ideological attack on leftist beliefs, goals, and ideas.[193] If that proved insufficient to repel leftists, or if other types of undesirables (e.g., rightists) were attracted to the movement, other means of keeping the movement "pure" would have to be found.

Rule (iii): The hard part would be the task of organizing people for practical action. We can't offer any formula or recipe for carrying out this task, but those who undertake such an effort will find their road less difficult if they apply the ideas and information provided in Chapter Four, which follows.

NOTES

1. Mao, p. 112.
2. Tocqueville, Vol. I, p. 172.
3. Huenefeld, p. 6.
4. "The propagandist must realize that neither rational arguments nor catchy slogans can, by themselves, do much to influence human behavior." NEB (2003), Vol. 26, "Propaganda," p. 175.
5. See Smelser, pp. 345n5, 356–57.
6. See NEB (2003), Vol. 16, "Collective Behavior," p. 563. There is undoubtedly a good deal of truth in what the *Britannica* says here about the blurring of a movement's goals over time. Nevertheless, the *Britannica's* statements are not entirely borne out by the examples discussed in the present chapter. We may suspect the *Britannica* of generalizing too broadly.
7. NEB (2003), Vol. 29, "War, Theory and Conduct of," p. 649.
8. Trotsky, Vol. Three, p. 179.
9. See Currey, p. 344; W.S. Randall, pp. 215, 250, 262.
10. E.g., McCullough, pp. 102, 163.
11. Ibid., pp. 374–381, 397–98. W.S. Randall, pp. 480–83. Chernow, pp. 227–239, 241, 243–44, 261–68.
12. W.S. Randall, p. 512.
13. Ibid., p. 201.
14. In the early days of the Republic the members of the Electoral College that chose the President were not necessarily elected by the people; in many states they were appointed by the state legislatures. Ibid., p. 544. NEB (2003), Vol. 29, "United States of America," p. 223. And, until 1913, Senators too were appointed by state legislatures, not elected by the people. *Constitution of the United States*, Article I, Section 3; Amendment XVII. The state legislatures that made these

appointments were not elected in a fully democratic way, because in most states the right to vote was limited by property qualifications. See NEB (2003), Vol. 29, "United States of America," pp. 217, 223, 269, 277, 283, 299, 302. Also Haraszti, pp. 32–33; Trees, p. 7A.

15. See NEB (2003), Vol. 16, "Christianity," pp. 258, 262 for the moral rigorism of the earliest Christians and its gradual relaxation. At least until circa 100 AD and probably through most of the following century, the Christians were numerically insignificant. Harnack, p. 5. Freeman, p. 163. Harnack finds in the reign of the emperor Commodus (180–192 AD) a turning-point in the growth of Christianity, Harnack, pp. 27ff, and by 300 AD perhaps seven to ten percent of the population of the Roman Empire was Christian, Freeman, p. 215. During the first half of the fourth century Christianity grew explosively and became dominant in the Roman Empire because it was favored by the emperor Constantine. Ibid., pp. 219, 222, 225ff.

16. Exodus 22:25. The *King James Version* uses the term "usury"; the *New English Bible*, the *Revised English Bible*, and the *New International Version* do not. In all these versions, strictly speaking, the Bible prohibits the taking of interest only from "poor" or "needy" people. But apparently Jesus's injunction to "lend, hoping for nothing again" (i.e., nothing in return; Luke 6:35, *King James Version*) was assumed to bar *all* lending at interest. See Weber, p. 59n1; Bouwsma, p. 198. (But see Matthew 25:14–28; Luke 19:12–25.)

17. For this whole paragraph, see *World Book Encyclopedia*, 2011, Vol. 20, "Usury," p. 229; NEB (2003), Vol. 12, "usury," p. 216; Pirenne, pp. 251–52&n4; Bouwsma, pp. 198, 202–03; Weber, pp. 56–58n1. Circa 200–250 AD, certain Christian widows were lending money at extortionate rates of interest. Harnack, p. 131. For money-lending at interest by Christians in later centuries, see, e.g., Pirenne, loc. cit.; Runciman, pp. 304–05; D. Jones, p. 384.

18. E.g., Matthew 6:19–24, 19:21–24; Luke 6:20–25, 12:15–21.

19. Acts 4:34–35 (*King James Version*). See also Acts 2:44–45, 4:32. Many scholars reject altogether the historical value of Acts, Freeman, p. 40, though Harnack, e.g., p.116, apparently does not. Whatever the value of Acts, it is sufficiently clear that the early Christians were *supposed to* despise wealth and hold their property in common. E.g., Freeman, pp. 163–64, 165.

20. Freeman, p. 115. Harnack, p. 26&n3. James 2:1–16, 4:13–16, 5:1–3. The Epistle of James was written not later than the early 2nd century AD. Freeman, p. 105.

21. Freeman, passim, e.g., pp. 169, 187–88 (greed), 211–12 (greed, failure to help starving brethren), 261, 266 (greed). Harnack, passim, e.g., pp. 21, 23, 25–32, 39 (greed), 106–08 (greed, total decadence), 116. It's true on the other hand that early Christians did more for the poor than the pagans had done, according to Freeman, pp. 267–69 (referring to early fifth century), but this could have been

motivated in part by the fact that taking care of the poor helped to ensure the stability of the existing structure of society, see ibid. Presumably for that reason, the church "was used by the state" to care for the poor. Ibid, p. 225. The State gave some bishops "grain supplies to hand out to the poor." Ibid., p. 228. Moreover, it's not clear that concern for poor people can be attributed specifically to *Christian* doctrines, since support for the poor is an important part of classical Judaism and of Islam (Neusner & Chilton, pp. 38–39, 74–77), and no doubt of many other religions. This writer is aware of no reason to believe that Christianity has done better in this respect than other major religions have. For whatever it's worth: "Whereas the principle of official support for the poor and assignment of work to the unemployed had been systematically developed under the Stuart monarchs [of England], especially during the regime of [William] Laud [Archbishop of Canterbury] under Charles I, the battle-cry of the Puritans was, 'Giving alms is no charity'. . . ." Weber, p. 177n3.

22. See, e.g., Freeman, p. 225 ("megachurch complexes").

23. In the *King James Version*, Exodus 20:13 says "Thou shalt not kill," and Mark 10:19 and Luke 18:20 both make Jesus say, "Do not kill," but Matthew 19:18 quotes Jesus as saying, "Thou shalt do no murder." However, in each of these four verses the more modern translations cited in note 16, above, use the word "murder" instead of "kill." Jesus must at least have considered it justifiable to kill in self-defense, for when he advised his disciples to carry swords (Luke 22:36) he surely did not mean these to serve as mere decoration. Arguments that Jesus's reference to "swords" was not literal but allegorical make no sense in the context; one would have to believe that the references to money, coats, shoes, purse, scrip, garment, etc. (Matthew 10:9–10; Mark 6:8–9; Luke 10:4, 22:35–36) likewise were not literal but part of the allegory. And if one believes that much, then one cannot assume that *anything* in the Gospels is to be taken literally. The meaning of the Gospels would be up for grabs—no one could know what Jesus meant by anything he said; hence, a fortiori, Jesus's teachings could not have guided human behavior. Of course, if we were to assume that Jesus really was a pacifist, then that would strengthen our argument that his teachings have been ineffective—in view of the amount of violence that the Christian world has seen since his time.

24. See, e.g., Elias, pp. 162–65, 171.

25. This discussion of Christianity has been conducted at a naïve level inasmuch as it assumes that Jesus actually said what the Gospels report him to have said. This simplified treatment has been necessary for the sake of brevity. For clarification, see Appendix Six.

26. See NEB (2003), Vol. 23, "Marx and Marxism," pp. 533–34.

27. See, e.g., ibid., pp. 539–542.

28. "When people speak of ideas that revolutionize society, they do but express the fact that within the old society the elements of a new one have been

created, and that the dissolution of the old ideas keeps even pace with the dissolution of the old conditions of existence." Marx & Engels, Chapt. II, p. 91.

29. "Propaganda that aims to induce major changes is certain to take great amounts of time, resources, patience, and indirection, except in times of revolutionary crisis when old beliefs have been shattered...." NEB (2003), Vol. 26, "Propaganda," p. 176.

30. M.F. Lee, pp. 119, 136.

31. The story is told by M.F. Lee.

32. Among the early Bolsheviks, Krasin and Bogdanov were essentially adventurers. Ulam, pp. 90, 95, 101–02. (Bogdanov, alias Alexander Malinovsky, should not be confused with Roman Malinovsky.) During the 1930s, in the opinion of Otto Bauer, there were many adventurers in the German anti-Nazi movement. Rothfels, pp. 64–65. See Packer, p. 62 (suggesting that many Muslim jihadists are motivated by a "sense of adventure").

33. See NEB (2003), Micropaedia articles on Anthony, Susan B.; Bajer, Fredrik; Blatch, Harriot Eaton Stanton; Braun, Lily; Catt, Carrie Chapman; Gage, Matilda Joslyn; Garrison, William Lloyd; Grimké, Sarah (Moore) and Angelina (Emily); Mott, Lucretia; Phillips, Wendell; Rankin, Jeanette; Stanton, Elizabeth Cady; Stone, Lucy; Truth, Sojourner; Woodhull, Victoria. Also Vol. 9, "prostitution," p. 737.

34. See ibid., articles on Anthony, Catt, and Gage, plus: Vol. 9, "Pankhurst, Emmeline," pp. 115–16; Vol. 19, "Feminism," p. 160 ("the feminist movement... became focused on a single issue, woman suffrage..."). It's true that beginning in the late 1890s "radical feminists challenged the single-minded focus on suffrage," ibid., p. 161, but still it seems clear that at least from about 1870 until the 1920s, woman suffrage was overwhelmingly the dominant goal of the feminist movement in England and America.

35. See ISAIF, ¶ 83.

36. "In connection with the Italian fascist movement, Rossi remarked that by the beginning of 1922, the movement was sufficiently successful to provide various advantages to members— 'uniform, arms, expeditions, subsidies, loot, flattery, and all the other advantages reserved to fascists.' Such attractions presumably would attract members on bases other than ideological commitment." Smelser, p. 357n1, quoting Rossi, p. 180.

37. Trotsky, Vol. Two, pp. 309–310. As ordinarily used, the term "opportunist" refers to an individual who makes use of an opportunity to advance his own personal interests without regard to moral or political principles. However, in Marxist-Leninist theory the "opportunists" were not unscrupulous individualists but socialists who focused on immediate goals of socialism, such as improving the economic status of the workers, rather than on the revolutionary goal of transforming society as a whole. See, e.g.: Stalin, *History of the Communist Party*,

passim; in particular, first chapter, Section 3, p. 30 (the " 'Economists'... were the first group of compromisers and opportunists..."). Lenin, "What Is to Be Done?," Chapt. I, Part D ("fashionable preaching of opportunism"); in Christman, p. 69. Selznick, passim, e.g., p. 308 ("Opportunism is a readiness to adapt to situations that offer immediate rewards without weighing the consequences of such adaptation for the ultimate character of the group."). Stalin eventually came to use "opportunist" as a general term of abuse for anyone whom he wanted to denounce. See his *History of the Communist Party*, seventh chapter, Section 2, p. 261 (accusing Kamenev, Zinoviev, etc. of opportunism); Conclusion, pp. 483–84. Trotsky, loc. cit., describes the "fundamental quality of opportunism" as "submission to the existing powers." Whatever Trotsky's exact meaning may have been, it can hardly be open to doubt that a great many of those who came flooding into the Bolshevik Party upon its attainment of power must have been opportunists in the ordinary, individualistic sense of the word.

38. Sampson, p. xxv.

39. See, e.g., P.P. Read, pp. 58–60. The progressive corruption (in our sense of the word) of early Christianity with the growth of its own power can be seen from Freeman, e.g., pp. 187–88, 201, 225 (especially), 237, 253, 261, 266, 269, 270, and Harnack, e.g., pp. 39, 106–08, 136, 241, 256, 259–260, 287.

40. NEB (2003), Vol. 12, "Uthman ibn Affan," p. 219; Vol. 22, "The Islamic World," pp. 110–11.

41. See R. Zakaria, e.g., pp. 59, 282–83, 296.

42. La Botz, pp. 43–63, 127. See also NEB (2003), Vol. 6, "Institutional Revolutionary Party," p. 333, and Vol. 24, "Mexico," pp. 48–49; Agustín, both volumes, entire. The revolutionary stage of Mexico's "revolutionary" party ended in 1940. See note 166, below.

43. Hoffer, § 116.

44. Ibid., § 7, quoting Hitler, p. 105.

45. Mao, pp. 362–63.

46. Ibid., p. 475.

47. *The Economist*, June 25, 2011, p. 14 ("Although the decision by these young careerists to sign up [for Communist Party membership] shows the party's clout, they have very different ambitions from those of the old ideologues."). Ibid., "Special Report" on China: The general impression one gets from this Special Report is that China's politicians are doing what politicians everywhere do—jockeying for personal power and advantage—and that they use the old Maoist ideology only as a tool for that purpose.

48. E.g., *The Economist*, April 2, 2011, p. 34, and April 23, 2011, p. 74; Folger, p. 145; *USA Today*, Sept. 3, 2014, p. 7A.

49. W.S. Randall, p. 357.

50. NEB (2003), Vol. 29, "United States of America," pp. 216–18.

51. McCullough, pp. 504–06, 536, 577.

52. Buckley, p. 7A ("But just how long did republican virtue persist in America? By 1829... there wasn't too much left."). See NEB (2003), Vol. 29, "United States of America," pp. 221, 223–24; McCullough, p. 398.

53. This according to La Botz, p. 66. But Agustín, Vol. 2, p. 121, claims that the civil society movement already existed in "nascent" form in 1976.

54. La Botz, pp. 66, 81.

55. Ibid., p. 72.

56. Ibid., p. 70.

57. Ibid., p. 81.

58. See ibid., p. 234.

59. Ibid., p. 78.

60. Ibid., p. 80.

61. Ibid., p. 232.

62. *USA Today*, July 3, 2018, p. 3A.

63. Ibid., July 3–4, 2012, p. 2A.

64. Ibid.

65. Ibid., May 10, 2017, p. 5A.

66. *The Week*, March 21, 2008, p. 11. See also ibid., March 13, 2009, p. 16; Caputo, pp. 62–69; Padgett & Grillo, pp. 30–33.

67. Foer, p. 45 (López Obrador "has embraced a more business-friendly persona"). *USA Today*, July 3, 2018, p. 3A ("he has tried to reassure financial markets").

68. See note 166, below. Of course, even if López Obrador's efforts to help the poor should prove highly successful, that would not represent an achievement of the civil society movement, which—it appears—has long since ceased to be a political force in Mexico.

69. *USA Today*, May 10, 2018, p. 5A; July 3, 2018, p. 3A.

70. See: note 66, above, and *USA Today*, Feb. 6, 2013, pp. 1A, 5A; Oct. 9, 2014, p. 7A; Oct. 20, 2014, p. 7A; Nov. 22, 2014, p. 8A; May 10, 2018, p. 5A. Also note 21 to Chapter Two, and Hayes, p. 3A (last paragraph).

71. See ISAIF, ¶ 100; note 66, above; Agustín (the entire work); La Botz (the entire work); *USA Today*, May 10, 2017, p. 5A, May 10, 2018, p. 5A, and July 3, 2018, p. 3A. Guillermoprieto, pp. 87–88, describes briefly the unofficial—and, from the point of view of legal governmental structures, corrupt—system of *caciques*, or local bosses, that plays a large part in governing Mexico. Guillermoprieto's description dates from 1994, but this writer knows of no reason to think that the system has changed since then. See also La Botz, p. 28.

72. The nascent civil-society movement apparently made some contribution to the sequence of events in 1976 that led to the lifting of government restrictions on freedom of the press in Mexico, Agustín, Vol. 2, pp. 119–121, and if

one wanted to stretch a point one might possibly list this as a success on the part of the movement. But by 2008 government restrictions had been replaced with those imposed by the drug gangs, which exercise censorship through the simple expedient of murdering any journalists who offend them. See note 70, above.

73. See NEB (2003), Vol. 19, "Feminism," p. 160.

74. See note 34, above.

75. NEB (2003), Vol. 12, "women's movement," pp. 734–35; Vol. 19, "Feminism," pp. 161–62.

76. See both the articles cited in note 75.

77. The reason for the trend is that this type of equality is to the advantage of the technological system. See Kaczynski, "The System's Neatest Trick."

78. Kee, pp. 24–25.

79. Ibid., pp. 101–09, 114–121, 126, 128–29, 151.

80. Ibid., p. 204.

81. Ibid., p. 179.

82. Ibid., pp. 181–82.

83. Ibid., pp. 181, 186.

84. Ibid.

85. Ibid., pp. 181–86.

86. Catholic Emancipation received only a "pained and angry" royal assent. Ibid., pp. 185–86. The king "had to be bullied" into approving the measure. NEB (2003), Vol. 29, "United Kingdom," p. 80.

87. Kee, pp. 184–85. Churchill, pp. 27–30.

88. Kee, pp. 152, 193, 201, 227.

89. Ibid., pp. 193–242.

90. Ibid., p. 246.

91. Ibid., p. 257.

92. Ibid., p. 261.

93. Ibid., pp. 264–67.

94. Ibid., pp. 270, 304–05.

95. Ibid., pp. 305–06.

96. Ibid.

97. Ibid., pp. 308–310, 315–320.

98. Ibid., pp. 335–340.

99. Ibid., pp. 351–564, especially p. 391.

100. See ibid., pp. 352–53.

101. Ibid., pp. 351–470, especially p. 352.

102. Ibid., pp. 548–709. Those who are familiar with Irish history may feel that the first half of this paragraph represents a serious oversimplification, but, assuming that Kee's history is not misleading, and notwithstanding the expected objections of Irish nationalists (who are not in a position to make an unbiased

judgment), I think this passage comes close enough to the truth for present purposes. I'm not writing a textbook of history. I'm using Irish history to illustrate certain points, and to do this *briefly* I have to paint with a very broad brush. Similar remarks apply to many of the other historical examples cited throughout this book.

103. Ibid., pp. 719, 726.

104. Ibid., p. 728.

105. Ibid., pp. 728n*, 730, 732–745.

106. Ibid., pp. 748–49. However, some diehards continued to maintain an illegal organization calling itself the Irish Republican Army. Ibid.

107. Ibid., pp. 748–751.

108. The nationalists' victory was incomplete inasmuch as the six counties comprising Northern Ireland remained part of the United Kingdom. This fact is not important at the present point in our discussion, but we will have occasion to return to it further on.

109. See ibid., pp. 389–390.

110. Ibid., pp. 353–54, 368–376.

111. See ibid., pp. 750–51.

112. See ibid., pp. 732–33, 752.

113. Ibid., p. 303 (Stephens's "political thought contained obvious traces of revolutionary socialist thinking"); p. 334 ("a republic... which shall secure to all the intrinsic value of their labour"); p. 751.

114. Ibid., p. 446. NEB (2003), Vol. 21, "Ireland," p. 1004 ("cultural revivalism became an inspiration to the Irish nationalist struggle of the early decades of the 20th century").

115. Ibid., p. 1001.

116. See McCaffrey & Eaton (referring to 2002 and the years immediately preceding), pp. 5, 120, 203, 219 (Burger King in Dublin). On the other hand, it must be admitted that Ireland has been rather slow in adopting modern culture. Ibid., pp. 21, 23, 120.

117. In an obvious attempt to escape temporarily from the modern world, some Irish play at reinacting Bronze Age customs—as they imagine them to have been. Ibid., pp. 44–47. Compare similar games played elsewhere in Europe.

118. Kee, p. 751.

119. See NEB (2003), Vol. 15, "Calvin and Calvinism," p. 436; Vol. 19, "Geneva," p. 743; Vol. 26, "Protestantism," p. 212, and "Rousseau," p. 939.

120. During the 1970s the Secretary General of the Communist Party of Spain wrote that the State that had developed in the Soviet Union was neither one "that could be considered a *workers' democracy*" nor "the State that Lenin imagined." Carrillo, pp. 201, 202. Carrillo also pointed out that, in the Soviet Union, "the bureaucratic stratum... decides and resolves at a higher level than the working class

and even at a higher level than the party, which, as a whole, is subordinate to the bureaucratic stratum." Ibid., pp. 207–08. Thus, the Soviet Union was a dictatorship neither of the proletariat nor of the party of the proletariat, but of the bureaucracy.

121. It's not entirely clear to what extent Putin actually monopolizes power in Russia, but, whatever the exact distribution of power may be, no one claims that Russia as of 2018 is a functioning democracy.

122. Tella, Germani, Graciarena et al., p. 266n15, quoting Fluharty, quoting in turn García.

123. See Kaczynski, Letter to David Skrbina: Nov. 23, 2004, Part IV.C.

124. NEB (2003), Vol. 29, "Western Africa," p. 841.

125. Ibid., Vol. 17, "Eastern Africa," p. 810.

126. Ibid., p. 803.

127. See note 22 to Chapter Two, and: NEB (2003), Vol. 17, "Eastern Africa," p. 798. *National Geographic*, Sept. 2005, p. 15. *Denver Post*, Feb. 26, 2009, p. 11A. *Time*, Aug. 23, 2010, p. 19. *USA Today*, Aug. 6, 2009, p. 7A; Aug. 7, 2017, p. 3A; Aug. 10, 2017, p. 3A.

128. S.A. Reid (the entire article). *USA Today*, Jan. 18, 2017, p. 5A. Ibid., Jan. 20–22, 2017, p. 10A, "TROOPS ENTER GAMBIA TO FORCE OUT JAMMEH." Yahya Jammeh was the dictator.

129. "For Lenin, organization was an indispensable adjunct to ideology." Selznick, p. 8. See Lenin, "What Is to Be Done?," Chapt. IV, Parts C, D, E; in Christman, pp. 147–48, 151–52, 157.

130. Kee, pp. 21–27. The "Whiteboys" apparently took their name from the white shirts they wore. Ibid., p. 24.

131. Ibid., pp. 24, 26, 27.

132. Ibid., pp. 44, 57, 68–69, 73.

133. Ibid., pp. 57, 59, 61, 68–69, 73, 126, 151.

134. See ibid., pp. 101–09, 114–121, 128–29.

135. Trotsky, Vol. One, pp. 136–152, especially p. 152. For the role of ideas in the French Revolution, see Haraszti, p. 22, citing opinion of M. Roustan.

136. Kee, pp. 450–611.

137. Ibid., pp. 595–742.

138. NEB (2003), Vol. 20, "Germany," pp. 89–90; Vol. 23, "Luther," p. 310. Dorpalen, pp. 113–14, 117, 119&n49.

139. Dorpalen, p. 113.

140. Ibid., pp. 114–121&nn44, 49. NEB (2003), Vol. 10, "Schmalkaldic League," p. 527; Vol. 20, "Germany," pp. 88–90; Vol. 23, "Luther," pp. 310–11; Vol. 26, "Protestantism," pp. 208–211, 213.

141. NEB (2003), Vol. 20, "Germany," p. 83.

142. See Elias, e.g., pp. 166–171. Cf. Graham & Gurr, Chapt. 12, by Roger Lane.

143. On learned helplessness see Seligman.

144. Trotsky, Vol. Two, p. 309.

145. Lenin, "What Is to Be Done?," Chapter IV, Part E; in Christman, p. 162.

146. Stalin, *History of the Communist Party*, second chapter, Section 3, pp. 60–61.

147. Ibid., Section 4, p. 68.

148. Mao, pp. 143–44, 172, 175, 258.

149. NEB (2003), Vol. 2, "Brook Farm," p. 549.

150. Smelser, pp. 356–57, quoting Noyes, p. 653.

151. Smelser, p. 357n1.

152. Selznick, pp. 24, 60.

153. Ibid., passim, e.g., pp. 120–21, 132–33, 178, 216, 221.

154. Ibid., pp. 21–36, 177.

155. NEB (2003), Vol. 12, "Woodhull, Victoria," p. 743. Buhle & Sullivan, pp. 36–37 ("Woodhull fled to England, married wealthily and renounced her former radical ideas.").

156. Our historian states explicitly that O'Connell had his later Repeal Association firmly under his own control. Kee, p. 193. O'Connell founded the Catholic Association (for Catholic Emancipation), and the Catholic Association was superbly disciplined, ibid., pp. 179–186, so it is probable in view of O'Connell's immense prestige that he had the Catholic Association under his own control to the extent that it is possible for one man to control an organization of that size.

157. Ibid.

158. Ibid., p. 182.

159. Ibid., p. 456.

160. See ibid., p. 450.

161. This statement slightly simplifies the actual situation. A relatively "moderate" Sinn Fein leader like Arthur Griffith was willing at least to contemplate conceding something to those Irishmen who wanted less than total independence. Ibid., pp. 451, 720. But Griffith personally believed that nothing less than total independence would be sufficient. Ibid., p. 451. Throughout Kee's account of the Irish War of Independence, his application of the term "moderates" to Griffith's faction evidently refers only to a difference of opinion within the movement about the means by which and the time at which total independence was to be achieved. There seems to have been general agreement within the movement that total independence was the ultimate goal. E.g., ibid., pp. 609 ("it was known that Sinn Fein stood for 'total independence'"), 626.

162. Ibid., pp. 606, 608, 610, 611, 621–22, 630, 641, 647–48, 651, 652, 654, 661, 680, 711, 732, 733.

163. Ibid., pp. 606, 613–14, 641, 654, 661, 662, 732.

164. See ibid., p. 730. Once hostilities against the British were suspended,

there was a split in the Irish movement owing largely to "the removal of the need for what had often been unnatural unanimity"—unanimity enforced by the requirement of a common front against the enemy in a shooting war.

165. NEB (2003), Vol. 6, "Institutional Revolutionary Party," p. 333. But the *Britannica's* statement may represent an unduly generous view of the motives of Mexico's revolutionary leaders other than Lázaro Cárdenas. Compare the detailed account of Tannenbaum, pp. 198–224.

166. La Botz, pp. 43–63. Agustín, Vol. 1, pp. 7–19, 49–52, 56–57, 68, 155–57. President López Mateos (1958–1964) made some gestures toward reviving the program of agrarian reform, but these were mostly ineffectual. Ibid., pp. 173, 196, 269. Presidents Echeverría Álvarez (1970–1976) and López Portillo (1976–1982) likewise attempted to renew agrarian reform. Their efforts were probably sincere (though La Botz, p. 128, may not think so) but, again, mostly ineffectual. Agustín, Vol. 2, pp. 20–23, 98–101, 146, 166, 171, 174, 229–232, 279, 289. For the coup de grace by Salinas de Gortari, see La Botz, p. 118.

167. A few individual women have gotten rich by writing feministic books, but they have been able to write these books independently of their membership in feminist organizations, if indeed they have been members of any such organizations.

168. Kee, pp. 726–27, 733–34.

169. Ibid., pp. 728&n*, 730–33, 748–49.

170. Ibid., pp. 748–751.

171. Ibid., e.g., pp. 610–12.

172. Ibid., pp. 750–51.

173. Ibid., e.g., pp. 592–93, 721, 745, 748.

174. See ibid., p. 749, and NEB (2003), Vol. 10, "Sinn Fein," p. 837.

175. Kee, pp. 713, 747, 749, 751, 752.

176. See note 174.

177. See NEB (2003), Vol. 26, "Protestantism," pp. 206–214.

178. See Dorpalen, pp. 108, 113, 121, 124, 125&n65.

179. Skrbina, pp. 275–78. According to Dr. Skrbina (personal communication to the author), Glendinning's article originally appeared in *Utne Reader*, March/April 1990. I have not seen the original article, and am relying solely on Dr. Skrbina's anthology.

180. Skrbina, pp. 221–230. The "essay" is actually a section of Naess, pp. 92–103.

181. Naess, p. 97. Skrbina, p. 225.

182. Naess, p. 98. Skrbina, p. 226.

183. Naess, p. 99, Table 4.1. Skrbina, p. 226.

184. Naess, p. 98. Skrbina, p. 226.

185. "The objectives of the deep ecological movement do not imply any

depreciation of technology or industry. . . ." Naess, p. 102. Skrbina, p. 229.

186. This writer has read but little of the work of Glendinning or of Naess, and it's possible that elsewhere in their writings they may have remedied to some extent the deficiencies we've noted here, by (for example) formulating more precise goals or acknowledging the need for an organization oriented toward practical action. For present purposes, however, this is not very important, because our interest is not in Glendinning or Naess personally but in the whole genre of literature that they represent. And the deficiencies we've noted in the works of Glendinning and Naess here discussed are characteristic of the genre.

187. See, e.g., ISAIF, ¶¶ 99–104, 111–12, and Kaczynski, Letter to David Skrbina: March 17, 2005.

188. Kaczynski, Letter to David Skrbina: April 5, 2005, Part II, and Extract from Letter to A.O., June 30, 2004.

189. See ISAIF, ¶¶ 207–212.

190. This was the opinion of, for example, the late distinguished astronomer Fred Hoyle (Hoyle, p. 62). The argument is that, due to the exhaustion of such natural resources as readily accessible deposits of coal, oil, and high-grade metallic ores, there could not be a new Industrial Revolution; consequently, there could never again be a technologically advanced society. Unfortunately, I can't agree with this. I think it's all too possible that a technologically advanced society might be developed without "coal, oil, and high-grade metallic ores," especially since there would be a vast amount of scrap metal left over from the previous technological society. But it's certainly true that the development of a technologically advanced society would be much slower and more difficult the second time around due to the lack of coal, oil, etc.

191. ISAIF, ¶¶ 6–32, 213–230.

192. See Kaczynski, Preface to the First and Second Editions, point 3, and the case of Judi Bari in the discussion of Postulate 3, above.

193. As in ISAIF, ¶¶ 6–32, 213–230; Kaczynski, "The System's Neatest Trick;" and (in the 2010 Feral House edition of Kaczynski) "The Truth About Primitive Life."

CHAPTER FOUR

Strategic Guidelines for an Anti-Tech Movement

> Force is the final arbiter, vigorous intervention is the keynote, and victory goes to those who have the courage and the discipline to see things through to the end. Such a view is characteristic of groups which seek to catapult themselves out of obscurity into history when, as it seems to them, all the forces of society are arrayed in opposition.
>
> — Philip Selznick[1]

1. No specific route to victory for an anti-tech movement can be laid out in advance. The movement will have to wait for opportunities that in due course will enable it to bring about the collapse of the technological system. The exact nature of the opportunities and the time of their arrival will in general be unpredictable, so the movement will have to prepare itself for successful exploitation on short notice of any and all such opportunities.

First, the movement must build its own internal sources of power. It will have to create a strong, cohesive organization consisting of individuals who are absolutely committed to the elimination of the technological system. *Numbers* will be a secondary consideration. A numerically small organization built of high-quality personnel will be far more effective than a much larger organization in which the majority of members are of mediocre quality.[2] The organization will have to develop its understanding of the dynamics of social movements so that it will recognize opportunities when they arrive and will know how to exploit them.

Second, the movement must build power in relation to its social environment. It must win respect for its ideas, its vigor, its effectiveness. If it is widely feared and hated, so much the better; but it must earn for itself a reputation as the purest and most uncompromisingly revolutionary of all oppositional movements. Thus it will be the movement to which many individuals will turn upon the arrival of a severe crisis in which people have become desperate and have lost all respect for and all confidence in the existing form of society.

Third, to help pave the way for this loss of respect and confidence, the movement should do what it can to undermine people's faith in the technological system. This is likely to be the lightest of the movement's burdens, because much of the work will be done without any effort on the part of the movement. For one thing, the system's own failures will help to undermine confidence in it. For another, the spoken and written words of disenchanted intellectuals, especially those concerned with environmental issues, will act (and are already acting) to break down people's confidence in the existing social order. Very few of these intellectuals are potential revolutionaries,[3] therefore an anti-tech movement should not support them directly. But the movement can promote the decline of confidence in the existing social order by calling attention to the pervasiveness and the irremediable character of the system's failures and by making the system look weak or vulnerable whenever possible.[4]

In this chapter we will try to fill in some of the details of the picture that is roughly sketched in the foregoing paragraphs.

2. Revolutions almost never are successfully planned out long in advance of their actual occurrence. This is merely one instance of the principle that specific historical events are, in general, unpredictable.[5] Irving Horowitz correctly observed that revolutions are carried out either without a previous program of action, or even in direct violation of such a program,[6] and Herbert Matthews noted that "of all the revolutionary leaders of modern times, only Hitler outlined his program and stuck to it."[7] Revolutionaries have to proceed by trial and error, and by grasping (usually unforeseen) opportunities as they arise.[8] As Lenin put it: "We often have to grope our way along... . Who could ever make a gigantic revolution, knowing in advance how to carry it through to the end?"[9] In January 1917, Lenin did not believe that any kind of revolution would be possible in Russia during his own lifetime.[10] He was able to make the Bolsheviks masters of Russia only because he had the acumen to recognize and exploit the unexpected opportunity presented by the February 1917 insurrection in St. Petersburg.[11]

3. Major opportunities, however, may be a long time in coming; the revolutionary movement may have to lie in wait for them.[12] This doesn't

mean that the movement can afford to relax and take it easy. On the contrary, while it is waiting the movement must remain hard at work, not only to build its strength so that it will be able to take full advantage of opportunities when they arrive, but also because an inactive movement will die or shrink to an apathetic rump. If a movement's members are not kept occupied with purposeful work, most will lose interest and drift away.[13]

Another reason why the movement must remain active is that it is not enough for revolutionaries to wait passively for opportunities; the opportunities may have to be created in part by the revolutionaries themselves. Some serious failure of the existing social order will probably have to occur independently of anything the revolutionaries can do, but whether such a failure is severe enough to provide an opportunity for overthrow of the system may depend on previous revolutionary activity. In Russia, for example, the underlying weakness of the tsarist regime was not caused by revolutionaries. But the opportunity for revolution was based on the regime's defeat in World War I, and revolutionary activity may have contributed to that defeat, for "[i]n no other belligerent country were political conflicts waged as intensively during the war as in Russia, preventing the effective mobilization of the rear."[14] Later, it was the spontaneous and unexpected insurrection of the workers of St. Petersburg that gave the Bolsheviks their great opportunity, and that insurrection probably would have been no more than a disorganized and ineffective outburst of frustration if the Bolsheviks had not previously indoctrinated the workers with Marxist ideas,[15] thus providing them with a theory and an ideal that made it possible for their insurrection to be purposeful, organized, and effective.

4. From section 2, above, it follows that a revolutionary movement has to be prepared to respond successfully to the unexpected. If a program of action is to cover any appreciable span of time, the movement must not be committed to it in such a way that the program cannot be altered or discarded as unforeseen developments may require. In other words, the movement must maintain *flexibility*.

Students of military tactics and strategy have long recognized the importance of flexibility.[16] Lenin demanded "tactical flexibility" in revolutionary work,[17] and Trotsky attributed the power of the Bolsheviks to the fact that they had "always united revolutionary implacableness with the greatest flexibility."[18] Mao Zedong wrote:

> [I]n the practice of... changing society, men's original ideas, theories, plans or programmes are seldom realized without any alteration. ... [I]deas, theories, plans or programmes are usually altered partially and sometimes even wholly, because of the discovery of unforeseen circumstances in the course of practice. That is to say, it does happen that the original ideas, theories, plans or programmes fail to correspond with reality either in whole or in part and are wholly or partially incorrect. In many instances, failures have to be corrected many times before errors in knowledge can be corrected and... the anticipated results can be achieved in practice. ...
>
> ...[T]rue revolutionary leaders must not only be good at correcting their ideas, theories, plans or programmes when errors are discovered, ... but... they must ensure that the proposed new revolutionary tasks and new working programmes correspond to the new changes in the situation.[19]

This is one way of describing the need for flexibility.

5. As argued in Chapter Three, the single ultimate goal of a revolutionary movement today must be the total collapse of the worldwide technological system.[20] One of this writer's correspondents has suggested that, because of the acute physical danger and hardship to which everyone would be exposed following a collapse of the technological system, a movement that takes such a collapse as its goal will be resisted by the overwhelming majority of the world's population and therefore will be unable to accomplish anything.

Undoubtedly, if you held a referendum today on the question of whether the system should be made to collapse, ninety percent, at the very least, of the inhabitants of industrialized countries would vote "no." Even in a crisis situation in which people had lost all respect for and all confidence in the system, it may well be that a majority, though a much smaller one, would still vote against total collapse. But the assumption that this would be a serious obstacle to revolution is based on what we may call the "democratic fallacy": the notion that the number of people favoring one side or another determines the outcome of social struggles as it determines the outcome of democratic elections. Actually the outcome of social struggles is determined not primarily by numbers but by the dynamics of social movements.

6. It goes without saying that the real revolutionaries—the members of the deeply committed cadre that forms the core of the movement—will

be prepared to accept any amount of hardship and the greatest risk, or even a certainty, of death in the service of their cause. We need only think of the early Christian martyrs; of Al Qaeda, the Taliban, and the Islamic suicide bombers; or of the assassins of the Russian Revolution. After a Social Revolutionary named Kalyaev assassinated a Russian grand duke in 1905, the duke's wife visited him in prison and told him: "Repent… and I will beg the sovereign to give you your life." Kalyaev replied: "No! I do not repent. I must die for my deed and I will. … My death will be more useful to my cause than [the grand duke's] death."[21]

Later, in 1918, when Fanny Kaplan put two bullets into Lenin, she surely realized that she would pay with her life.[22] Similarly, when Charlotte Corday assassinated Jean-Paul Marat during the French Revolution, she must have known that she would face the guillotine.[23] The extreme Irish nationalists who carried out the uprising of April 1916 certainly knew that they were taking desperate risks, and a small minority among them were intentionally seeking martyrdom. Many of those who were subsequently executed "expressed in their last words… confidence that their deaths were a sort of triumph."[24]

7. But it's not only a tiny minority of hard-core revolutionaries who will accept suffering and the gravest risks in the service of what they regard as critically important goals. Many ordinary people become heroes and show astonishing courage when there is a severe disruption of their society or an acute threat to their most cherished values, or when they are inspired by what seems to them a noble purpose.

It has been said that "man is capable of standing superhuman suffering if only he feels sure that there is some point and purpose to it."[25] This statement has been confirmed by experience, not only in the histories of the French, Russian, and other revolutions, but in many other situations as well. In World War II, for instance, the Russians never lost their will to resist in the face of the death, destruction, and savage cruelties inflicted on them by the German invaders.[26] For that matter, the morale of the German civilian population was never broken by the horrific Allied bombing campaigns that reduced many of their cities to rubble and sometimes killed tens of thousands of people in a single operation.[27] The Allied air-crews who carried out bombing and other missions in disputed air-space over Europe suffered in turn a frightful rate of attrition. For example, of the American pilots who undertook missions over German-occupied Poland during World

War II, about three out of four were killed.[28] Yet the survivors kept flying. Meanwhile, on the ground, many infantrymen suffered equal danger and far greater physical hardship, but they too continued to fight.[29]

Most of the civilians in the examples of the foregoing paragraph did not suffer hardship or danger voluntarily; they showed their courage merely by continuing to function well under the atrocious conditions imposed on them by circumstances beyond their control. Some of the military men no doubt volunteered for service, but probably many of these at the time they volunteered failed to appreciate fully what they were getting into. This was certainly the case with Audie Murphy, the most decorated American soldier of World War II, who was totally naïve about war when he enlisted.[30] Yet there are abundant examples of people—not just a tiny minority of hard-core revolutionaries, but large numbers of more-or-less ordinary people—who in critical situations have voluntarily chosen to take desperate risks, with what we can assume was full knowledge of what they were risking, in the service of a cause or in fulfillment of what they believed to be their duty. In 1922, when the Irish War of Independence had gone on long enough so that its desperate and bloody character was unmistakable, there was still no shortage of recruits, "new eager young warriors anxious to emulate their elders."[31] Nor does there seem to have been any shortage of recruits to the French and Polish resistance movements during World War II. These risked not only death, as the Irish did, but excruciating torture as well. Charles de Gaulle's personal representative with the French Resistance, Jean Moulin, was captured and tortured to death by the Gestapo,[32] yet he never cracked, never gave up his secrets.[33] "In 1941 Free France had sent Captain Scamaroni to [Corsica] with a mission to prepare action there. ... Unfortunately, our valiant delegate had fallen into the hands of the Italians.... Tortured horribly, Scamaroni had died to keep his secrets."[34]

Even for causes in which they have no personal stake, some people will risk death, and worse. Thus thousands of non-Jewish Poles participated in efforts to save Jews from the Nazis. In helping Jews the Poles risked death not only for themselves but for their families as well.[35] A Polish woman named Irena Sendler, credited with helping to save 2,500 Jewish children, "was captured by the Nazis in 1943 and tortured but refused to say who her co-conspirators were. During one session her captors broke her feet and legs... ." She survived only because her comrades in the Resistance bribed a Gestapo officer to help her escape.[36]

It should be noted, too, that whether they are hard-core revolutionaries or ordinary people, whether they assume their risks voluntarily or involuntarily, many of those who go through extreme danger or hardship for what they believe to be worthy purposes experience deep fulfillment from their "heroic" activities. They may even enjoy them. A former inmate of a German prisoner-of-war camp in World War II wrote of his unsuccessful and eventually successful attempts to escape:

> I feel I have quaffed deeply of the intoxicating cup of excitement.... I can think of no sport that is the peer of escape, where freedom, life, and loved ones are the prize of victory, and death the possible though by no means inevitable price of failure.[37]

As World War II drew to a close:

> Apart from the Communist leaders, who aimed at a definite goal, the resistance fighters as a whole were somewhat disoriented. As the enemy withdrew... they had been tempted, like Goethe's Faust, to say to the moment, 'Stay, you are so splendid!'... Nostalgia was upon them. Especially since these ardent and adventurous men had experienced, in the height of danger, the somber attractions of the clandestine struggle, which they would not renounce.[38]

Much more recently, with the arrival of peace in Northern Ireland, the withdrawal of these same "somber attractions" seems to have had a decidedly negative effect on the youth of that country. In 2009 a journalist reported his conversations with a Catholic priest, Father Aidan Troy:

> [T]he suicide rate among Belfast's youth has risen sharply since the Troubles ended, largely because, the priest believes, the sense of camaraderie and shared struggle provided by the paramilitary groups has been replaced by ennui and despair. 'So many young people get into drinking and drugs early on,' Troy says.[39]

Celia Sánchez, who had been a revolutionary guerrillera in Cuba, reminisced in 1965 about the dangers and hardships she had gone through with Fidel Castro's band in Sierra Maestra: "Ah, but those were the best times, weren't they? We were all so very happy then. *Really*. We will never be so happy again, will we? *Never*... ."[40]

In an otherwise rather maudlin article, an American veteran of the Iraq war conceded that his return to civilian life had its drawbacks: "I miss that daily sense of purpose, survive or die, that simply can't be replicated in everyday existence."[41]

8. The purpose of the foregoing examples is not to glorify danger, suffering, or warfare. Their purpose is to show that people—even the members of modern technological society, who in normal times are oriented primarily toward security and comfort—will not necessarily choose the easiest road, or the one that seems least dangerous in the short term, when their society is in turmoil, when they are desperate, angry, or horrified at the turn that events are taking, or when it no longer seems possible to maintain their habitual pattern of living. Under such circumstances many will choose a heroic course of action, even a course that subjects themselves and their loved ones to the greatest risks and hardships—if only there are leaders who can energize them, organize them, and give them a sense of purpose. It will be the task of revolutionaries to provide that kind of leadership when the system arrives at a crisis.

At such a time, if the revolutionaries have done and continue to do their work well, they should be able to attract wide support in spite of all the risks and hardships that the revolutionary program entails. This is not to say that the revolutionaries will succeed in winning the support of a majority of the population. It's much more likely that they will be able to organize and lead only a fairly small minority. But "it is not always the physical majority that is decisive; rather, it is superiority of moral force that tips the political balance." (Simón Bolívar).[42] In the event of a sufficiently serious failure of the existing social order the vast majority of the population will lose all respect for it and all confidence in it, hence will make no effective effort to defend it. Alinsky stated the case very clearly when he wrote that the "time is... ripe for revolution" when

> masses of our people have reached the point of disillusionment with past ways and values. They don't know what will work but they do know that the prevailing system is self-defeating, frustrating, and hopeless. They won't act for change but won't strongly oppose those who do.[43]

Under these circumstances a great many people will have become hopeless, apathetic, and passive, while most of the rest will be concerned only to save their own skins and those of their loved ones. It is to be expected that the existing power-structure will be in disarray, disoriented, and riven by internal conflict, so that it will do a poor job of organizing and leading any small minority that may still be motivated to defend the system. If, therefore, the revolutionaries act effectively to inspire, organize, and lead their own minority, they will hold the decisive share of power.

9. A failure of the existing social order may not always be needed to provide revolutionaries with an opportunity. It's not clear that there was any grave failure of the social order in Ireland prior to the revolution of 1916–1922; certainly the British authorities against whom the revolution was directed were by no means in disarray or otherwise weak. Yet the revolution did occur.[44] Ordinarily, however, an opportunity for revolution depends on some serious failure of the existing social order.

The Reformation was possible only because the corruption of the Catholic Church led many people to lose their respect for it. The revolutions of the early 19th century that won independence for Spain's American colonies probably would not have occurred if the weakness of the Spanish monarchy had not been demonstrated through its defeat by Napoleon and in other ways. The Chinese revolution of 1911 was largely a result of the repeated humiliations inflicted on China by the Western powers and Japan, against which the Manchu (or Qing, Ch'ing) Dynasty was unable to defend itself. The Russian revolutionaries were given their opportunity by the ignominious military defeats of the Tsarist regime in World War I. In Germany, the Nazis were a minor party up to the onset of the Great Depression; Hitler was able to seize power only because the German government was weak and unable to deal with the economic crisis.[45]

In each of the foregoing examples there undoubtedly was a broadly generalized loss of respect for the prevailing social order, and in the last two cases it is probably safe to say that there was widespread anger and desperation on the part of some people, hopelessness on the part of others. In today's world a prerequisite for revolution most likely will be a situation of the latter type, involving widespread anger, desperation, and hopelessness. Revolutionaries need to be capable of making use of such a situation.

To illustrate with a hypothetical example, let's suppose that in the coming decades the replacement of human workers by increasingly advanced technology will lead to severe, chronic unemployment throughout the technologically developed part of the world.[46] This will not necessarily produce a crisis serious enough to endanger the existence of the system, for people will tend to react to chronic unemployment with apathy, passivity, and hopelessness. There will be anger, too, which may lead to riots like those seen in Spain and Greece in 2011–12,[47] but these poorly organized, largely purposeless outbursts of frustration (really manifestations of hopelessness) accomplished little or nothing.

Compare this ineffectual rioting with the "Arab Spring" revolution in Egypt (2011), in which intelligent leadership harnessed people's anger and made it into a tool for the extraction of major concessions from the power-structure. The Egyptian revolution failed in the end, but for present purposes that is irrelevant. The point here is simply that skillful revolutionary leaders can harness people's anger and frustration and turn it to useful purposes.

Anti-tech revolutionaries, of course, can't be satisfied with extracting concessions from the power-structure; they have to bring it down altogether. If, as we've hypothesized, there is severe, long-lasting unemployment throughout the technologically advanced part of the world, most of those who still have jobs will be frightened and will have lost their respect for the system, but will be motivated only to hold on to their jobs as long as they can. The unemployed will be either apathetic and hopeless, or angry and desperate, or both. If there is widespread rioting it will put the power-structure under stress, but will not seriously threaten its survival. Well-prepared revolutionaries, however, should be capable of organization and leadership that will put people's anger and desperation to work, not in mere rioting, but for purposeful action. From our present standpoint the nature of the purposeful action can only be a matter for conjecture, but, just to take a speculative example, the revolutionaries might extract concessions from the power-structure as the Egyptians did, with the difference that the concessions would have to go far enough so that they would deeply humiliate the power-structure. This could be expected to break down the morale of the individuals comprising the power-structure and lead to sharp internal divisions and conflicts within the power-structure, throwing it into disarray. Once this stage had been reached, the prospects for the overthrow of the power-structure would be excellent.

But let's remember that the foregoing scenario represents a purely hypothetical route to revolution that we've offered only for illustrative purposes. Revolution may take a very different route in reality.

10. It is important to recognize that a successful revolutionary movement may start out as a tiny and despised group of "crackpots" who are taken seriously by no one but themselves. The movement may remain insignificant and powerless for many years before it finds its opportunity and achieves success. "Beliefs that are potentially revolutionary may exist temporally long before strain arises to activate these beliefs as determinants of a value-oriented movement; revolutionary organizations may lie in wait for conditions of conduciveness, upon which they then capitalize."[48]

In 1847 Karl Marx and Friedrich Engels were just a couple of eccentrics who prepared the Communist Manifesto for an obscure group called the Communist League, which had only a few hundred members and soon dissolved.[49] In Ireland, nationalist ideas were kept alive for several decades only by a minuscule minority of extremists who had very little support among the general population until the uprising of April 1916 reactivated the revolutionary process.[50]

Fidel Castro said, "I began a revolution with eighty-two men. If I had to do it again, I would do it with ten or fifteen and absolute faith."[51] Castro actually started his revolution with only about a dozen men, because three days after he landed in Cuba with his eighty-two they were attacked by the forces of the dictator, Batista; nearly all were killed or captured, and no more than twelve, or possibly fifteen,[52] were left to carry on the struggle in the Sierra Maestra. Even at its peak two years later the guerrilla band amounted to only about 800 men, as against Batista's army of 30,000.[53] Yet Castro won.

Such a victory of course could not be a purely military one, nor was it achieved by Castro's guerrilleros alone. Castro's victory was primarily a political one, and was possible only because the Cuban people had no respect for or confidence in the Batista regime. The dictator was politically incompetent and unable to retain the loyalty even of his own army, which proved itself decidedly reluctant to fight the rebels. And Batista was really overthrown by a coalition of forces, of which Castro's guerrilla band was not the only important component. What enabled Castro to prevail over the other elements of the coalition and emerge as master of Cuba was his skill as a politician, propagandist, and organizer. While his military action played an indispensable role, it did so mainly through its political and psychological effect.[54]

The point to be emphasized here, though, is that when Castro, leading his tiny band of a dozen men, looked up at the Sierra Maestra and said, "Now Batista will be defeated!,"[55] most people would have thought him mad. Yet Batista was indeed defeated and Castro did take control of Cuba.

In Russia at the beginning of the 20th century the revolutionaries comprised an insignificant minority and were regarded as "cranks."[56] The Russian Social Democratic Labor Party, of which the Bolsheviks formed a part, consisted of only a few hundred individuals.[57] According to Lenin:

> Prior to January 22... 1905, the revolutionary party of Russia consisted of a small handful of people, and the reformists of those days... derisively called us a 'sect'. ... Within a few months, however, the picture completely changed. The hundreds of revolutionary Social Democrats 'suddenly' grew into thousands; the thousands became leaders of between two and three million proletarians... .[58]

The 1905 revolution was a failure, but it did help prepare the way for the successful revolution of 1917. Up to the latter year, nevertheless, the Bolsheviks remained weak. At the outbreak of World War I in 1914, three of the seven members of their St. Petersburg committee were police spies, and soon afterward the Bolsheviks' centralized organization was destroyed by the arrest of their delegates in the Duma (the Russian parliament).[59] On the very eve of the opening episode of the 1917 revolution the Bolshevik leaders were scattered in exile, and no one (except possibly the police) paid any attention to them.[60] But less than a year later they had made themselves masters of the vast Russian Empire—something like one-sixth of the world's land surface (discounting Antarctica).[61]

The Bolsheviks had prepared themselves long in advance of the outbreak of the revolution. They had built a cohesive cadre of professional revolutionists who were disciplined, purposeful, strongly motivated, well led, and reasonably unified. The Bolsheviks were effective organizers, and, because they understood better than anyone else the dynamics of social movements, they formulated policies that proved to be successful. Their chief rivals, the far more numerous Social Revolutionaries, were deficient in these qualities. "[W]hereas the agitation of the Mensheviks and Social Revolutionaries was scattered, self-contradictory and oftenest of all evasive, the agitation of the Bolsheviks was distinguished by its concentrated and well thought-out character."[62] Trotsky describes how, in one county, three or four Bolsheviks were sufficient to prevail over the much larger but relatively timid Social Revolutionary organization.[63] "The lack of correspondence between the technical resources of the Bolsheviks and their relative political weight [found] its expression in the small number of members of the party compared to the colossal growth of its influence."[64]

Meanwhile, the "bourgeois-democratic" reformists (Kerensky et al.) were not even in the running, because they lacked unity and concentrated purpose and seem to have had no conception of what was and what was not possible in a time of passionate upheaval such as that which gripped Russia in 1917. As for the defenders of the old Tsarist order, to the extent that there were any left in Russia they were in total disarray and psychologically defeated. Consequently, the Bolsheviks were able to overwhelm all their adversaries and make themselves the dominant political force in Russia.

All this doesn't necessarily mean that the Bolsheviks had the support—much less the *active* support—of a majority of Russians. The support of the peasants was shaky at best, and existed only when the Bolsheviks were (temporarily) giving them what they wanted.[65] But once the Bolsheviks had seized power in October [66] 1917, the only *organized* and *effective* resistance to them originated outside Russia with the numerous émigrés who opposed the revolution. These assembled counterrevolutionary armies and, supported by several foreign powers, invaded Russia with the intention of ousting the Bolsheviks. During the ensuing Civil War of 1918–1920: "The rate of desertions in the Red Army was unusually high: Trotsky instituted a veritable reign of terror to prevent defections, including placing in the rear of the troops machine-gun detachments with instructions to shoot retreating units."[67] But obviously the Bolsheviks couldn't have maintained their control over a disaffected majority without the loyal support of at least a substantial minority; those machine-gunners wouldn't have been willing to shoot down their fellow soldiers on orders from Trotsky if they hadn't been committed to the Bolshevik cause. The Bolsheviks moreover had their minority well organized and disciplined;[68] consequently they prevailed over the invaders, who were poorly organized.[69]

It's important to notice that the crucial events of the Russian Revolution took place in St. Petersburg. This was true of the spontaneous insurrection of February 1917 and also of the Bolsheviks' seizure of power the following October. Thus the Bolsheviks were able to concentrate their efforts on a single city; once they had won in St. Petersburg the rest of the country was relatively easy.[70] This shows how victory at the single most critical point can provide a basis for the assumption of power throughout an entire society—a further reason why it is possible for a numerically small revolutionary movement to prevail.

11. To summarize, the expected pattern for a revolution against the technological system will be something like the following:

A. A small movement, a cohesive cadre of committed, hard-core revolutionaries, will build its internal strength by developing its own organization and discipline. This movement should have branches in several of the world's most important nations or groups of nations; say, the United States, China, Western Europe, and one or more of Russia, Latin America, and India. In each country, the movement will prepare the way for revolution by disseminating ideas—ideas that will be chosen for their soundness and not for their popularity. The movement will take pains to demonstrate the most uncompromising revolutionary integrity, and will strive to prove itself the most effective of all the factions opposed to the existing system.

B. A large minority of the general population will recognize that the revolutionaries' ideas have some merit. But this minority will reject the revolutionaries' solutions, if only through reluctance to change familiar ways of living or as a result of cowardice or apathy.

C. Eventually there will arrive a crisis, or a failure of the system serious enough to enable the revolutionaries to create a crisis, in which it will no longer be possible to carry on with familiar ways of living, and in which the system's ability to provide for people's physical and psychological needs will be impaired to such an extent that most people will lose all respect for and all confidence in the existing social order, while many individuals will become desperate or angry. Their desperation and anger will soon degenerate into despair and apathy—unless the revolutionaries are able to step in at that point and inspire them with a sense of purpose, organize them, and channel their fear, desperation, and anger into practical action. Because these people will be desperate or angry and because they will have been energized by the revolutionaries, the risk to themselves, however great it may be, will not deter them from striving to bring down the system.

D. Even so, the revolutionary movement will probably be able to gain the active support only of some fairly small minority of the population. But the great majority will be either hopeless and apathetic or else motivated merely to save their own skins, so they will not act to defend the system.

E. The established authorities meanwhile will be disoriented, frightened, or discouraged, and therefore incapable of organizing an effective defense. Consequently, power will be in the hands of the revolutionaries.

F. By the time revolutionaries have taken power in one nation—for example, the United States—globalization will have proceeded even farther than it has today, and nations will be even more interdependent than they are now.[71] Consequently, when revolutionaries have brought the technological system to an abrupt halt in the United States, the economy of the entire world will be severely disrupted and the acute crisis that results will give the anti-tech revolutionaries of all nations the opportunity that they need.

G. It is extremely important to realize that *when the moment for decisive action arrives* (as at C, above) *the revolutionaries must recognize it, and then must press forward without any hesitation, vacillation, doubts, or scruples to the achievement of their ultimate goal.* Hesitation or vacillation would throw the movement into disarray and would confuse and discourage its members. (We will return to this point in a moment.)

The pattern we have just outlined is a very broad and general one that can accommodate a wide variety of routes to revolutionary success. Even so, given the unpredictability of historical events, it is impossible to know for certain whether the route that a revolutionary movement will actually take will fit within the pattern we've described. But the pattern is an entirely plausible one, and it provides an answer to those who think the system is too big and strong ever to be overthrown. Moreover, the preparatory work that we have briefly indicated above, at A, will be appropriate for almost any route to revolution that a movement might take in reality.

12. Let's return to point G, above: that the revolutionaries must avoid all hesitation or vacillation when the moment for decisive action arrives. The leaders of the movement must be astute enough to recognize the arrival of that moment. Trotsky claims that in a revolutionary situation there is a particular interval of time, limited to a few weeks or at most a few months, during which a society is primed for insurrection. Any attempt to bring about an insurrection must be undertaken during that interval or the opportunity will be lost.[72] So says Trotsky, and we may accept that this is true as a general rule (though of course all such rules have exceptions). Trotsky was speaking only of insurrections, but it should be obvious that a

similar rule applies to many other kinds of revolutionary actions: One can hope to carry them out successfully only when circumstances are favorable for them, and since circumstances change rapidly when a society is in crisis one must act at the right time; to act too soon or too late will lead to failure.

Here we are concerned mainly with the right moment to begin organizing on a mass basis for the final push toward the overthrow of the existing social order (as at C, above), a push that may or may not involve one or more insurrections but almost certainly will not consist merely of a single insurrection. The critical interval of time may be difficult to identify. "Lenin... greatly feared excessive caution, ... a letting slip of one of those historic occasions which are decades in preparation."[73] On the other hand, if the revolutionaries act prematurely they may suffer a disastrous defeat. Only an assiduous study of history and of revolutionary theory, with careful and thoughtful observation of current events, can develop the judgment necessary for recognition of the critical interval during which the push toward consummation of the revolution can be successfully initiated.

But let's assume that the revolutionaries have correctly noted the arrival of the time to begin organizing on a mass basis for the final push. Once that stage has been reached, certain guidelines need to be taken into consideration.

Alinsky maintains that the organizers of a mass movement must "act in terms of specific resolutions and answers, of definiteness and certainty. To do otherwise would be to stifle organization and action, for what the organizer accepts as uncertainty would be seen by [the people he is organizing] as a terrifying chaos."[74] Trotsky warns against "indecisiveness": "The party of revolution dare not waver—no more than a surgeon dare who has plunged a knife into a sick body."[75] Here Trotsky refers to the final stage of a revolutionary process, when the existing social order is in a state of crisis and the revolutionaries are aiming directly at its overthrow. Throughout this stage there is a need to maintain *momentum*: Alinsky emphasizes that a mass movement has to remain constantly in action, avoid defeats, and keep its adversaries under unremitting pressure.[76] Trotsky says that a revolutionary process can continue only "so long as the swing of the movement does not run into objective obstacles. When it does, there begins a reaction: disappointments of the different layers of the revolutionary class, growth of indifferentism and therewith a strengthening of the position of the counter-revolutionary forces."[77]

However, the rule that momentum should be maintained is not unqualified: Revolutionaries should not, for the sake of momentum, undertake a major action prematurely. In July 1917 the Bolsheviks intentionally aborted an insurrection in St. Petersburg because they judged that the time was not ripe for it. Their action temporarily checked the momentum of the revolutionary process and led to a severe setback for the Bolsheviks, but it averted the utterly disastrous setback that would have ensued if the insurrection had actually been attempted.[78] Nothing in this is inconsistent with the rule that revolutionaries must act decisively and without vacillation: The Bolsheviks did indeed act decisively to abort an insurrection that they had done nothing to instigate and that they knew was untimely.

Alinsky stresses the importance of avoiding moral ambiguity. The organizers of a mass movement need to delineate issues in black and white: Their own cause must be pure, noble, unequivocally good, while their adversaries represent nothing but evil.[79] All of the movement's actions are automatically presumed to be fully justified, for any vacillation on moral or humanitarian grounds would be as fatal as vacillation on any other grounds. The fact that vacillation on moral or humanitarian grounds was likely to be fatal in any life-and-death conflict[80] was understood by some of our most admired statesmen and soldiers—those who led the Western democracies when they were locked in struggles for survival. E.g., Lincoln and Grant during the U.S. Civil War, or Churchill and Roosevelt during World War II.

Similarly, it is a fatal error to delay action, or to act timidly, in order to avoid offending people. For example: The Bolsheviks and the Mensheviks were the two revolutionary parties derived from the split in the Russian Social Democratic Labor Party. In the period immediately following the St. Petersburg insurrection of February 1917, Trotsky says, "the official Social Democratic program was still... common to the Bolsheviks and the Mensheviks, [and] the practical tasks of the democratic revolution looked the same on paper to both parties." But, while the Bolsheviks promptly undertook radical measures, the Mensheviks temporized in order to avoid antagonizing the bourgeoisie and the liberals.[81] In general, according to Trotsky, the behavior of the "Compromisers" (= Menshevik and Social Revolutionary leaders[82]) was "evasive." "The Compromisers talked themselves out of difficulties; the Bolsheviks went to meet them."[83] The Compromisers' tactics would have been appropriate under normal circumstances in a functioning parliamentary democracy, but in a revolutionary situation those same tactics were sure losers. So of course it was the Bolsheviks who came out on top.

The remarks in the last four paragraphs are intended to provide general guidelines for hard-core revolutionaries to take into consideration in the process of acquiring and leading a mass following when the system moves into a state of crisis; it is the volatile mass that will be incapable of tolerating uncertainty, moral ambiguity, defeats, or periods of inactivity. During the earlier stages of the movement's life, while it is diligently and patiently preparing the way for revolution, the hard-core revolutionaries, the committed cadre, will have to be able to endure—up to a point—the uncertainties that will inevitably arise, as well as the long periods without spectacular activity and the tactical defeats that will occur. But once the revolutionary process has arrived at its final stage—the time of crisis during which the revolutionaries are pushing directly toward the overthrow of the system—the committed cadre must strive to eliminate *even within its own ranks* all uncertainties, hesitations, vacillations, doubts, and scruples. For one thing, such internal vacillations would inevitably be communicated to the revolutionaries' mass following. For another, at this critical time it will be especially important for the committed cadre to be capable of prompt, decisive, unified action, and such action will be rendered impossible by vacillations or disagreements within the cadre. If vacillations or disagreements are long continued, even the most deeply committed revolutionaries may lose heart.

In practice, of course, vacillations and disagreements will probably arise among the revolutionary leaders even during the final push toward overthrow of the system. The revolutionaries will need to resolve these conflicts quickly and completely, so that they can show unity in action and provide their mass following with consistent, unambiguous, decisive leadership. "The high temper of the Bolshevik party expressed itself not in an absence of disagreements, waverings, and even quakings, but in the fact that in the most difficult circumstances it gathered itself in good season by means of inner crises, and made good its opportunity to interfere decisively in the course of events."[84]

As always, the reader must remember that in the real world events are unpredictable. The preceding paragraphs provide only general guidelines, not rigid rules that can be applied mechanically. The guidelines may have to be modified to adapt them to the concrete situations that will arise in the practice of revolutionary politics.

13. One possible cause of hesitation on the part of revolutionaries needs to be addressed. Some time ago this writer received a letter from an individual who asked whether revolutionaries should strive to bring about the collapse of the technological system even though the chaos attendant on the collapse would entail an increased risk of nuclear war. The answer is that revolutionaries should not be deterred by such a risk.

First, the proliferation of nuclear weapons to unstable or irresponsible countries (such as Pakistan, North Korea, and Iran) continues and is unlikely to be permanently halted.[85] Consequently, the risk of nuclear war can only increase as long as the technological system survives, and the sooner the system collapses the less will be the risk of nuclear war in the long run.

Second, though many people assume that a major nuclear war would result in the extinction of the human race and of most species of mammals, that assumption is probably incorrect. Undoubtedly the consequences of such a war would be horrible, but serious students of these matters do not believe that most species of mammals would be completely wiped out or that the human race would disappear.[86]

Third, if nothing intervenes to prevent the technological system from proceeding to its logical conclusion, there is every reason to believe that the eventual result will be a planet uninhabitable for all of the more complex forms of life as we know them today. See Chapter Two, Part IV. So if we had to choose between a major nuclear war and the continued existence of the system, we would have to take nuclear war as the lesser evil.

Fourth, if we allow the defenders of the system to deter us with the threat of nuclear war or of any other dire consequences, then we may as well give up. A revolutionary movement can't be successful if it allows its pursuit of its objective to be limited by reservations or qualifications of any kind, for these can only lead to fatal hesitation at critical times. Revolutionaries must take their goal to be the collapse of the system *no matter what*. You have to make a decision: Is the elimination of the technological system worth all of the desperate risks and terrifying disasters that it will entail? If you don't have the courage to answer "yes" to that question, then you'd better quit whining about the evils and hardships of the modern world and just adapt yourself to them as best you can, because nothing short of the collapse of the system will ever get us off the road that we are on now.

14. In sections 12 and 13 we've offered some guidelines for revolutionary action to be taken upon the arrival of an acute crisis of the system. Remaining to be discussed is the long preparatory period during which the movement builds its strength for the final push toward revolution.

In a revolutionary situation—as we've pointed out already in section 1—victory is determined not primarily by numbers but by the dynamics of social movements. In section 10 we've seen examples of numerically tiny movements that have initiated successful revolutions. A small but well-organized,[87] unified, and deeply committed movement will have a far better chance of success than will a vastly larger movement that lacks these characteristics. In other words, quality is more important than quantity.[88] Consequently, while an organization is building its strength for a future revolution, it must strictly subordinate the goal of increasing its numbers to that of recruiting high-quality people who are capable of total commitment to the cause. Their commitment must be exclusive; they must have no competing loyalty to any other cause. Because the membership of the revolutionary organization has to be limited, as far as possible, to people of this type, selectiveness in recruitment is essential.[89]

15. If the goal of revolutionaries is the complete elimination of the technological society, then they must discard the values and the morality of that society and replace them with new values and a new morality designed to serve the purposes of revolution.[90] Trotsky put it this way:

> Bolshevism created the type of the authentic revolutionist who subordinates [his ideas and his moral judgments] to historic goals irreconcilable with contemporary society... . [T]he Bolshevik party created not only a political but a moral medium of its own, independent of bourgeois social opinion and implacably opposed to it. Only this permitted the Bolsheviks to overcome the waverings in their own ranks and reveal in action that courageous determination without which the October [Revolution] would have been impossible.[91]

Suitable recruits to the revolutionary movement will include only those who are prepared to abandon the old values and morality and adopt in their place the revolutionary values and morality. The revolutionary message needs to be addressed to and designed for, not the general public, but the small minority of people who have the potential to become committed members of the revolutionary organization.

16. It follows that the revolutionaries should never retreat from their extreme positions for the sake of popularity or to avoid offending the moral or other sensibilities of the general public.[92] If the revolutionary organization were to dilute its message or prevaricate in order to avoid offending people it would discourage its own members and lose their respect, weakening their commitment to the organization; it would lose the respect of the best kind of potential recruits while attracting many who were incapable of total commitment to the organization; and it would lose the respect of the general public. A revolutionary organization should seek not to be liked, but to be respected, and it should have no aversion to being hated and feared. Mao regarded hatred of a revolutionary organization as a sign that it was effective.[93] It is to such an organization that many people will turn in a time of crisis when they have lost all confidence in the existing social order and are desperate or angry.

17. Revolutionaries will not suddenly become effective agitators, propagandists, organizers and leaders at the moment when the system reaches a crisis. They will need to begin developing these abilities through practical experience long before the crisis arrives. In order to acquire such experience, revolutionaries will have to involve themselves in political efforts that are peripheral to the central issue of technology. For example, an anti-tech organization might join with other groups in addressing some environmental issue of special importance—though it will be necessary for the revolutionaries to make very clear that the environmental issue is a sideshow and that the long-term goal must be to eliminate the entire technological system.

In all such activities the revolutionary organization should strive to prove itself more determined and more effective than the other groups involved, for when a crisis arrives the organization will more readily acquire a mass following if it has already demonstrated its superior effectiveness. "[I]n the course of struggle... broad masses must learn from experience that we fight better than the others, that we see more clearly than the others, that we are more audacious and resolute."[94]

Another way revolutionaries can acquire practical experience will be through the publication of a newspaper or journal devoted to anti-tech work. Lenin wrote:

> A paper is not merely a collective propagandist and collective agitator, it is also a collective organizer.... With the aid of, and around a paper, there will automatically develop an organization that will be concerned, not only with local activities, but also with regular, general work; it will teach its members carefully to watch political events, to estimate their importance and their influence on the various sections of the population, and to devise suitable methods to influence these events through the revolutionary party. The mere technical problem of procuring a regular supply of material for the newspaper and its regular distribution will make it necessary to create a network of agents... who will be in close contact with each other... .[95]

Nowadays, of course, a newspaper or journal will likely be published not only in print but also on the Internet; or perhaps even on the Internet alone.

18. In order to be effective, a revolutionary organization must be capable of unity in action. As Fidel Castro put it: "No one can expect anything useful from an organization comprised of anarchic men who, at the first disagreement, seek their own road, breaking and destroying the machine." Consequently, Castro put great importance on discipline.[96]

Stalin stressed the need for "unity of will" and "absolute and complete unity of action on the part of all members of the Party." He set forth an admirable theory:

> [Unity] does not mean of course that there will never be any conflict of opinion within the Party. On the contrary, iron discipline does not preclude but presupposes criticism and conflicts of opinion within the Party. Least of all does it mean that this discipline must be 'blind' discipline. On the contrary iron discipline does not preclude but presupposes conscious and voluntary submission, for only conscious discipline can be truly iron discipline. But after a discussion has been closed, after criticism has run its course and a decision has been made, unity of will and unity of action become indispensable conditions without which Party unity and iron discipline in the Party are inconceivable.[97]

Needless to say, Stalin was concerned above all to maintain his own power, and consequently he never allowed the democratic aspect of the foregoing theory to be put into practice. But this need not prevent us from recognizing that the theory itself—that decisions are to be arrived at with free discussion and criticism throughout the organization, after which all members will be expected to obey the decisions that have been made whether or not they personally agree with them—is an excellent one for a revolutionary organization to follow.

Nelson Mandela would have agreed with Stalin's theory (though not, of course, with Stalin's practice), for he "believed passionately in democracy" within the African National Congress,[98] yet insisted on party discipline: Once a decision had been made by the organization, all members had to comply with it. "Having subjugated his own will to the movement, he was determined that others should do so too."[99]

But it has to be conceded that in practical terms the theory is not as democratic as it sounds. First, many decisions will need to be made quickly, with no time for discussion by the rank and file. The organization will have to have some sort of executive body that is empowered to make such decisions, and the rank and file will have to obey the decisions so made. Second, even when there is sufficient time, the organization can't be effective if many decisions are made by a simple head-count, so many votes on one side, so many on the other. However offensive it may be to our democratic sensibilities, the plain truth is that some individuals will have vastly more knowledge and experience relevant to the functioning of the organization than others will. Every member of the organization should be listened to, but the main responsibility for decision-making will have to rest with a relatively small group of leaders[100] comprising those members who are best informed and have the highest level of political and organizational skill. Thus, an effective revolutionary organization will require a significant measure of hierarchy and discipline.[101]

The so-called "democratic" countries in today's world are in reality governed by political parties. In even the most democratic of these parties, decisions are made primarily by a limited inner circle of leaders[102] who pay only as much attention as they think expedient to the opinions of the rank and file. A close approximation to true democracy can exist only in societies organized on a very small scale, such as the nomadic bands of African pygmies.[103] In any modern, large-scale society, a political organization that attempts to maintain a truly democratic internal structure will condemn itself to impotence.

19. Recognition of the importance of unity might lead to an erroneous conclusion, namely, that a revolutionary organization should never

split when there are disagreements over principles, strategy, or tactics. Of course, a faction shouldn't split from its parent organization for slight reasons or while there is a good prospect of resolving disagreements through discussion, or when there is an acute, immediate need to present a united front against adversaries. But an organization cannot be truly unified when there is within it a persistent, irreconcilable disagreement over a question of far-reaching importance. If such a disagreement develops among the members of a revolutionary organization, and if there is no apparent likelihood of resolving the disagreement within a reasonable time, it will usually be best if the dissident minority separates itself from the parent group. This will leave the parent group and the minority each with its own independent unity. If the minority is wrong it presumably will remain weak, while the parent group leads the revolution. On the other hand, if the minority's view is proven right through practice, then the minority can be expected to assume leadership when the time is ripe and leave its parent organization in the dust.

Lenin said, "We must not be afraid to be a minority,"[104] and he never hesitated to act accordingly when he was sure he was right. Trotsky makes clear that Lenin always insisted on pursuing his own line no matter what the rest of the Bolsheviks thought. Lenin preferred to be a member of a small minority that was *right* rather than compromise his views in order to get broader support.[105] Thus he and his Bolsheviks, though they constituted a minority within the Social Democratic Party, split from their rivals, the Mensheviks (effectively in 1903, formally in 1912) and took their own road.[106] Because their road turned out to be the right one, they eventually prevailed over the Mensheviks.

Again, at the outbreak of World War I in 1914, Lenin adopted and maintained an anti-war position and even advocated "transforming the imperialist war into civil war," despite the fact that he was supported in this only by his "closest comrades," who comprised "a minority within the group of anti-war Socialists, who, in turn, constituted a small minority of the international Socialist movement... ."[107] Lenin and his minority prevailed in the end because their judgment of the political situation had been better than that of other socialists.

When Lenin announced his "April Theses" in the spring of 1917 these were met with hostility by the other Bolshevik leaders, who thought he was "temporarily disorientated."[108] Lenin persisted, however, and in this case he did succeed after several weeks in bringing the rest of the party over to his view.[109] Much the same thing happened in October of that year when, at first against the opposition of the majority of the Bolshevik leaders but eventually with success, Lenin advocated the insurrection that put the Bolsheviks in control of Russia.[110]

Lenin won out in these conflicts only because his political judgment was better than that of his opponents. If his opponents had advocated more effective policies, they would have prevailed in the end and Lenin would have sunk into obscurity.

Lenin of course was a political genius, so he could afford to be confident to the point of arrogance in his political judgments. Those of us who are not equally gifted should be more cautious about risking a split in a revolutionary movement. Nevertheless, when it has become clear that there are deep and irreconcilable disagreements between different factions, it will generally be advisable for a movement to split.

20. A revolutionary movement needs to be self-confident. Alinsky, in explaining the techniques he had used throughout his long and successful career as a social and political activist, emphasized that a community organizer had to have confidence in himself[111] and had to instill confidence in the people he was organizing. As long as people lacked confidence in their own power to bring about great changes they remained passive and apathetic, but once they were imbued with a sense of their own power they could become energetic, active, and effective.[112] Trotsky noted the significance of the fact that the Bolsheviks "believed in their own truth and their victory."[113] The international communist movement—successor to the Bolsheviks—placed importance on "belief in the triumph of our cause."[114]

When Fidel Castro claimed that he could start a revolution with ten or fifteen men (see above, section 10), he added an important condition: His men had to have "absolute faith," presumably meaning absolute faith in their own eventual victory. The term "absolute faith" must be taken with a grain of salt. Given Marxism's claim to be "scientific" and the enormous prestige of science, it's not surprising that many Marxists of the 19th and the early 20th century had absolute faith in the eventual victory of the proletarian revolution. But nowadays well-informed people are more sophisticated, more skeptical. If you try to tell them that your movement is absolutely certain to achieve victory, you will attract only those who are either thoroughly irrational or extraordinarily naïve.

Castro, however, in speaking of "absolute faith," may have been referring not to a literal belief in the certainty of victory but to a psychological state: to buoyant self-confidence and a subjective sense of power—qualities that encourage people to exert themselves to the limit, to recover from repeated defeats, and to persist in the face of difficulties that less inspired individuals would see as insurmountable. This psychological state does not require an absolute certainty of success, but it does at least require a belief that one will have an excellent chance of success if only one works hard enough and long enough and shows sufficient energy, courage, willpower, skill, and determination.

Such a belief can be rationally sustained. Self-confidence tends to be self-justifying, in the sense that confidence that one can succeed tends to lead to *actual* success. A chief determinant, if not *the* chief determinant, of success for a revolutionary movement is its faith in itself. Faith leads to deep commitment; it inspires heroic efforts and persistence in the face of overwhelming difficulties. Given such faith and commitment, a movement may achieve things that no one thought possible. Above, section 10, we've given examples of tiny groups of seeming "cranks" who initiated successful revolutions against what appeared at the outset to be impossible odds. Numerous examples can be cited—we will cite some in a moment—of groups that eventually achieved victory only because they had the self-confidence to persist in the face of defeat and even when their situation seemed hopeless.

Conversely, when people lack confidence in their power to achieve things they will not in fact achieve anything difficult, because no one will exert himself to the limit when he has little hope that his efforts will be rewarded with any impressive result. For the same reason it is a serious mistake to set modest goals for a revolutionary movement on the ground that such goals are "realistic." Only a truly world-transforming goal can inspire people to accept hardship, risk, and sacrifice, and to put forth the extreme effort that will be necessary for the success of any real revolutionary movement in the world today.[115]

It follows that the goal a revolutionary movement sets itself must be nothing less than the total collapse of the technological system. The movement moreover must consistently insist that its chances of achieving that goal will be excellent if its members show a sufficient level of commitment, energy, courage, willpower, skill, and persistence.

21. An important note of clarification: The rule that a revolutionary movement should be self-confident refers to confidence in its ability to reach its ultimate goal—that of consummating the revolution. Overconfidence in carrying out particular projects or operations must be carefully guarded against, because overconfidence leads to carelessness and carelessness leads to failure. That's why Lenin habitually exaggerated the potential risks involved in any action and worked out his plans with meticulous care.[116] As Trotsky said, "one must be prudent to win the right to be bold."[117]

Prudence demands that one take care not to underestimate one's adversary. Underestimation of the adversary leads to overconfidence, thence to carelessness and defeat. In general, it is safer to overestimate one's adversary. Such was the policy of Lenin.[118] Mao emphasized that while one must have confidence in one's ability to defeat the enemy in the long run, one must never slacken one's efforts through overconfidence during the actual process of struggle:

> Comrade Mao Tsetung has repeatedly pointed out: strategically, with regard to the whole, revolutionaries must despise the enemy, dare to struggle against him and dare to seize victory; at the same time, tactically, with regard to each part, each specific struggle, they must take the enemy seriously, be prudent, carefully study and perfect the art of struggle. . . .[119]

In line with this, it should be understood that the rule that a revolutionary movement must have an ambitious, world-transforming goal refers only to the movement's ultimate goal. The movement's subsidiary goals—the goals that are steps on the way to the ultimate goal—should be prudently and carefully selected. Mao advised, "fight no battle you are not sure of winning."[120] Mao apparently was thinking primarily of a military situation, but whether in a military or in any other situation, his advice would be impractical if taken in a strictly literal sense. Seldom can one be really *sure* of success in any enterprise. However, in contemplating any project or action, revolutionaries should cautiously balance the advantages to be gained through success against the risk of defeat. Trotsky pointed out: "Every defeat... changes [the correlation of forces]... to the disadvantage of the vanquished, for the victor gains in self-confidence and the vanquished loses faith in himself."[121]

The hard core of a revolutionary movement needs to have the confidence, the commitment, and the psychological toughness to recover from repeated defeats and carry on in spite of them. But even the most deeply committed revolutionaries are, after all, human, and may be weakened by defeats or failures. Therefore one should risk a defeat or a failure only when there is a strong reason for doing so.

22. The negative effect of defeats will be mitigated if revolutionaries understand that, following a crushing defeat that seems to leave a group in a hopeless situation, a determined renewal of effort by whatever is left of the group very often leads to victory.

In a surprise attack at midwinter, 877–78, Danish Vikings seized control of Wessex, the country of the West Saxons. Believing that resistance was futile, the Saxons submitted to the invaders, but their king, Alfred, escaped with a few followers to the woods and moors of Somerset, and by Easter 878 he had established himself in a fort on an island in the Somerset marshes. At some point, either before or after reaching the marshes, Alfred collected a small army, and from the fort his men harassed the Danes with guerrilla attacks. About the middle of May Alfred summoned Saxon warriors from neighboring parts of Wessex and marched with them against the Danes, whom he defeated decisively at the Battle of Edington.[122] Alfred's "memory lived on through the Middle Ages and in legend as that of a king who won victory in apparently hopeless circumstances."[123]

Even more impressive is the case of Robert Bruce.[124] Toward the end of the 13th century, Edward I of England occupied Scotland and made it into something like an English colony. The Scots were restive under English rule, and in 1306 Robert Bruce, whose ancestry gave him a claim to the kingship, had himself inaugurated as King of Scotland. But within three months he was defeated in battle by the forces of Edward I and became a hunted fugitive, forced at times to survive under conditions of the greatest hardship.[125] At this stage his cause seemed hopeless. He had hardly any money or troops,[126] and the weakness of his position was "almost ludicrous."[127] Nevertheless, over the succeeding years Bruce waged a savage guerrilla campaign, gradually increasing the territory he controlled and the number of his followers until, in 1314, he defeated the English decisively at the Battle of Bannockburn. After that he reigned in effect as King of Scotland, though he did not secure English recognition of Scotland's independence until 1328. Bruce's rise from a hunted fugitive to ruler of an independent kingdom is seen by some as incredible,[128] but it does not look incredible to those who have noticed how often in history seemingly lost causes have eventually triumphed.

In the autumn of 1878, the Social Democratic movement in Germany was very nearly destroyed by the Socialist Law of October 19 of that year, which was enforced with extreme severity and had the effect of abolishing any "societies with 'social-democratic, socialistic, or communist' tendencies."[129] "As their foes were encouraged, many of the Social Democrats lost heart. ... [T]he movement nearly disintegrated completely."[130] But within a year some of the tougher and more persistent Social Democrats were publishing a paper in Switzerland and devising ways of smuggling it into Germany.[131] Meanwhile, other members of the movement developed legal and illegal subterfuges that enabled them to circumvent the Socialist Law and build a new organization for the party,[132] so that by the autumn of 1884 German Social Democracy was stronger than ever[133]—even though it was still illegal.

According to Mao, "in 1931... some comrades became proud and overweening. The result was [a]... serious error in the political line, which cost us about 90 percent of the revolutionary forces that we had built up with so much toil."[134] An editors' note explains:

> The erroneous 'Left' line dominated the Party for a particularly long time (four years) and brought extremely heavy losses, with disastrous consequences, to the Party and the revolution. A loss of 90 percent was inflicted on the Chinese Communist Party, the Chinese Red Army and its base areas... .[135]

But the Communists persisted in their efforts, rebounded from their defeats and, as we know, by 1949 had made themselves masters of China.

In South Africa during the early 1970s the ANC (African National Congress) seemed thoroughly defeated and almost defunct.[136] But what was left of the organization continued the struggle, with the result that the ANC eventually recovered its strength, made itself the dominant political force in South Africa, and subsequently became the ruling party of that country.

The Bolsheviks repeatedly recovered from severe defeats. When the Social Democrats of Russia (who included the Bolsheviks[137]) "helped to rouse antigovernment demonstrations" in 1905, their insurrection failed, and "they were arrested, imprisoned, or exiled."[138] To one who lived through those days it seemed that "[t]he revolution was dying. ... Darkness and despair had set in [among the intelligentsia]."[139] "But Lenin did not despair of success. ... For him there were lessons to be learned, new plans to be worked out, alternate methods of revolution to be considered."[140]

Again in 1914, at the outbreak of World War I, "the revolutionary movement died down. ... The revolutionary ideas were barely kept glowing in small and hushed circles. In the factories in those days nobody dared to call himself 'Bolshevik' for fear not only of arrest, but of a beating from the backward workers."[141] As we mentioned earlier (section 10), the Bolsheviks' centralized organization was destroyed at this time through the arrest of their delegates in the Duma. Nevertheless the Bolsheviks persisted, and following the February 1917 insurrection and the implementation of Lenin's "April Theses" they made themselves into an important force in the Russian revolutionary process.

However, as a result of the "July Days" (the abortive insurrection of July 1917;[142] see section 12, above) the Bolsheviks again suffered a severe setback,[143] one that would have been fatal to a less determined group.

> 'After the July Days,' writes V. Yakovleva, at that time a member of the Central Committee..., 'all reports from the localities described with one voice not only a sharp decline in the mood of the masses, but even a definite hostility to our party. In a good number of cases our speakers were beaten up. The membership fell off rapidly, and several organizations... even ceased to exist entirely.'... The efflux from the party in some cases reached such a scale that only after a new registration of members could the organization begin to live a proper life.[144]

We've emphasized that any major defeat is dangerous. But if a revolutionary organization has a hard core that is absolutely committed and determined, the organization in *some* cases may actually be strengthened by a defeat because its weaker members are weeded out: If they don't leave the organization, they at least reveal themselves by their wavering during the period of failures and difficulties. Thus the hard core is consolidated, because its members are clearly distinguished from the weaker members of the organization. Trotsky notes in reference to the July Days:

> This sharp turn in the mood of the masses produced an automatic, and moreover an unerring, selection within the cadres of the party. Those

> [Bolsheviks] who did not tremble in those days could be relied on absolutely in what was to come. They constituted a nucleus in the shops, in the factories, in the districts. On the eve of [the Bolshevik seizure of power in October 1917], in making appointments and allotting tasks, the organizers would glance round many a time calling to mind who bore himself how in the July Days.[145]

In this way the Bolsheviks drew an advantage from their July defeat when the time came for them to take control of Russia. But just a few months after their seizure of power they again came close to total defeat with the invasion of the "White" counterrevolutionaries and their Western allies:

> [T]he Bolsheviks were about to fall. It seemed a matter of days. Ruin surrounded them, from the Pacific and all across Siberia and the Urals, their power had collapsed. The Germans were in charge in the Ukraine, where a voluntary army was forming against the Bolsheviks, and the English were landing in the north. As was famine.[146]

In these circumstances, nothing but the unbreakable determination of the hard core of the Bolshevik Party enabled it to survive. But it did survive, and it retained its iron grip on Russia for more than sixty years thereafter.

This ability to bounce back from severe defeats is a trait that seems characteristic of successful revolutionary leaders. The trait is delineated with particular clarity in the case of Fidel Castro. Matthews emphasizes "Fidel's incorrigible optimism and fighting spirit"[147]:

> 'The most important feature of Fidel's character,' his brother Raúl said to me..., 'is that he will not accept defeat.'
>
> Every phase of his life, from childhood to the present, proves this point. ... Fidel never gave up; he never lost heart; he seems immune to discouragement and dismay.[148]

> Fidel Castro was like Lenin in having the gift of inspiring all those around him by his faith in himself and in what he was doing. ... [I]t showed up best in the worst and apparently most hopeless periods.[149]

23. In these pages we may seem to be making heroes of such men as Robert Bruce, Lenin, Mao, Castro, the extreme Irish nationalists, and so forth. Certainly the deeds of all these people were of heroic magnitude. But this doesn't mean that we should admire them as human beings, still less that we should respect their goals or their values. The Bolshevik/Communist leaders were committed technophiles,[150] and therefore the adversaries of those of us who believe that modern technology is pushing the world toward disaster. Robert Bruce may (or may not) have made some pretense of patriotic motives,[151] but in all probability his real motive was personal ambition[152]—he wanted to be king of Scotland—and in the service of that ambition he inflicted terrible cruelties not only on the English but even on some of his fellow Scotsmen.[153] In the twentieth century, as we pointed out in Chapter Three, there was no reason why Ireland needed to become independent of Britain.[154] It was solely in order to satisfy their own psychological needs that the Irish nationalists provoked the war of independence that brought suffering and death to so many of their countrymen, and the Irish are no better off today than they would have been if Ireland had remained part of the United Kingdom.

Here we've taken notice of some of the revolutionaries of the past only because we can learn something from their experience and their methods. If we've cited Communist leaders more often than others, we've done so not from any sympathy for Communism but only because the Communists, by and large, have been the most effective and successful revolutionaries of the 20th century.

24. Professional propagandists know that people usually accept only those new ideas that they are already predisposed to accept.[155] A revolutionary movement should try to identify the sectors of the population whose members are most likely to be predisposed to accept the revolutionary message, and should give special attention to those sectors in propagating its ideas and in its efforts at recruitment. Nevertheless, anti-tech ideas should be made known not only to the predisposed sectors but to the population at large. The rule that only predisposed people accept new ideas is not necessarily applicable "in times of revolutionary crisis when old beliefs have been shattered."[156] Thus, as we pointed out in section 8, when a severe crisis of the system arrives the revolutionary movement will have its opportunity to acquire a mass following; but a mass following will be more easily acquired if most people already have at least some superficial

acquaintance with anti-tech ideas. Moreover, even long before the arrival of a crisis and even in sectors where the revolutionaries cannot hope to win any active support, their message can promote discontent and disillusionment and thus help to set the stage for the arrival of the crisis. See in this chapter section 1, third point, and Alinsky as quoted in section 8.

25. A revolutionary movement must maintain clear lines of demarcation that separate it from other radical groups holding ideologies that to some extent resemble its own.[157] This is a corollary to the need for unity that we stressed in section 17: A social or political movement can't be unified if it has many members whose loyalty is divided between their own movement and some other. Moreover, a movement needs to have a clear and unmistakable identity of its own; this is necessary not only for the internal cohesion of the movement itself, but also so that outsiders will easily recognize the movement and will respect it (see section 1, second point, and section 16). In addition, the movement needs to keep itself strictly independent of all other groups. Dependence upon or too close a linkage with another group will prevent a revolutionary organization from acting in the interest of its own goals when these conflict with the goals of the other group.

One movement from which an anti-tech organization needs to separate itself definitively is that of the radical environmentalists; another is anarchoprimitivism. Most radical environmentalists do not contemplate the elimination of the entire technological system. An anti-tech organization can't afford to have members who are not sure they really want to eliminate modern technology, nor can it afford to be linked with a movement that holds an ambivalent position respecting technology. The anarchoprimitivists do want to eliminate modern technology, but other goals are at least equally important to them: gender equality, gay rights, animal liberation, etc.—the whole catalog of leftist issues.[158] Elsewhere we've explained why an anti-tech movement must emphatically distance itself from leftism.[159]

26. In its relations with rival radical groups, a revolutionary organization should avoid getting entangled in sterile, interminable wrangles over ideology. Such wrangles have been prevalent, for example, in anarchist circles. Some anarchists seem to spend most of their time and energy on theoretical squabbles with other anarchists and very little on efforts to bring about the social changes that they advocate. Neither side in these

disputes ever succeeds in persuading the other, and no one but the participants has any interest in the arguments offered.

Seldom indeed will you succeed in persuading your opponents in an ideological dispute. Therefore, in any such dispute, your arguments should be designed not to persuade your opponents but to influence undecided third parties who may hear or read the arguments. For this purpose you should state your case concisely, as clearly and convincingly as possible, and in a way that will make it interesting to third parties. Then do what you can to ensure that your arguments are widely heard or read. Address only the most important points and leave out the minor ones, for third parties will be interested only in the main lines of the arguments. Squabbles over arcane technical points are worse than a waste of time because third parties, if they read them at all, will probably view them with disdain and may compare you to the medieval theologians who quarreled over the number of angels who could dance on the point of a pin. Similar principles apply to debates with the defenders of the existing system, and with those who don't defend the system as it now exists but think it can be reformed.

When one is confronted with arguments that attack one's ideas or one's group one is strongly tempted to answer them, and the more unreasonable the arguments are, the stronger is the temptation to answer them. But before one gives in to this temptation one should ask what advantages, if any, one's answer can win for the revolutionary organization, and one should consider whether there are other ways of spending one's time and energy that will be more useful for revolutionary purposes than an answer to the offensive arguments would be.

The way to prevail over rival radical groups is not to argue with them but to outflank them: Focus on recruiting to *your* organization any suitable persons who are predisposed to reject modern technology but are undecided among the various factions. Show that your organization is more active and effective than other radical groups. This will bring more people over to your viewpoint than any amount of argument will do.

27. "[T]he most precious of all revolutionary qualities, loyalty, has its inevitable counterpart in treachery."[160] Members of any radical organization need to bear in mind at all times the likelihood that their group includes informers who will report their activities to law-enforcement or intelligence agencies, and they should remember that even individuals who are currently loyal may turn traitor at some later date.

From 1956 to 1971 the FBI implemented a program known as COINTELPRO that involved, among other things, the systematic infiltration of informers into groups that the FBI found politically objectionable.[161] COINTELPRO under that name has long since been discontinued but, needless to say, the FBI still uses similar methods today. In 2006, members of a group of eco-saboteurs were arrested with the help of an FBI informer who had infiltrated radical-environmentalist circles.[162] At about the same time, in a related operation, the FBI arrested the group that had been responsible for the spectacular eco-arson at Vail, Colorado in 1998. One of the group's members had turned traitor and helped the FBI to collect evidence; some of the others subsequently testified against their comrades in order to get shorter sentences for themselves.[163]

In South Africa the police used spies and informers with devastating effect against anti-apartheid activists, and some of the activists, when subjected to interrogation, gave information that helped the police to arrest their colleagues.[164] In Ireland, revolutionary groups were regularly infiltrated by government informers (though by 1919, under Michael Collins, the revolutionaries had turned the tables and developed a much better intelligence network than that of the government).[165] Fidel Castro's guerrilleros felt it necessary to execute many traitors whom they discovered in their ranks.[166] Of the members of Che Guevara's guerrilla band in Bolivia, some who were captured gave the authorities information about the members who were still free.[167] During the period in which the Social Democrats of Germany were outlawed (1878–1890), they established an "intelligence system" for the purpose of "sifting and analyzing raw information to uncover informers and agents provocateur[s],"[168] but this did not entirely protect them against infiltration by police agents.[169] Even one of the delegates to the Social Democrats' secret congress at Wyden Castle in Switzerland (August 1880) was "in the pay of the Berlin Police President."[170] In Russia, the revolutionary movement was thoroughly infiltrated with spies and informers.[171] The Social Revolutionaries' "Combat Organization" was headed for a time by a police agent,[172] and according to Trotsky, as noted in section 10, above, three of the seven members of the Bolsheviks' St. Petersburg committee in 1914 were police agents.[173] A prominent Bolshevik named Roman Malinovsky, who was the party's spokesman in the Duma and played a critical role in the founding of *Pravda*, later turned out to be a police agent. Even after it should have been evident that Malinovsky was a spy, Lenin refused to believe it.[174]

The pattern is consistent and the lesson is clear: A radical group can never safely assume that its plans or its activities are unknown to the government. Thus, a legal revolutionary organization is well advised to remain exactly that: strictly legal.[175] Any sort of dabbling in illegal activities is extremely dangerous.

28. It is important to study the history and the methods of earlier social and political movements and the techniques developed by successful leaders of such movements. It is a serious mistake to reject out of hand the techniques and the theories of revolutionaries or activists of the past merely because their goals were incompatible with anti-tech goals or because they were leftists or reformists. It's true that many of their methods must be rejected as unsuitable for use by an anti-tech organization today, and of their other methods many must be modified to adapt them to such use. Neither history nor the principles laid down by past leaders will provide formulas or recipes for success that can be applied in cookbook fashion. But they provide *ideas*, of which some may lead to methods that are suitable for anti-tech use while others may call our attention to dangers or stumbling-blocks that we need to avoid.

Mao emphasized not only the importance of learning from the experience of the past as recorded in history, but also that theories derived from past experience were often incomplete and needed to be corrected through further experience. Similarly, principles of action found to be valid in other contexts might not be applicable to the concrete situations arising in the development of a given revolution. Consequently, from among such principles revolutionaries needed to sort out what was useful for their purposes from what was useless, discard the useless, and modify the useful to adapt it to their own needs.[176]

It takes hard work to study the history and the methods of past movements and to sort out the useful from the useless. But if you fail to learn from the past then you condemn yourself to learning everything all over again, by trial and error. This is a slow, halting, and difficult process. A good deal of trial and error will be necessary anyway, but the number of trials needed and the number of errors committed will be greatly reduced if you put out the effort demanded by a careful study of earlier movements and their methods. A refusal to make this effort will seriously diminish your chances of success.

This writer has had no opportunity to study more than a few of the works of history, political science, sociology, and revolutionary theory that may be relevant to the anti-tech enterprise. Worthy of careful attention are the works of Alinsky, Selznick, Smelser, and Trotsky that appear in our List of Works Cited. But there is a vast amount of other relevant literature that deserves to be explored; for example, the literature of the academic field known as "Organizational Behavior," and the works of Lenin to the extent that they deal with revolutionary strategy and tactics (his ideological hokum is merely of historical interest). Thorough library research will reveal an unending series of other relevant works. It is worth repeating that this literature will provide no recipes for action that can be applied mechanically. It will provide *ideas*, some of which can be applied, with suitable modifications, to the purposes of an anti-tech organization.

29. Let's illustrate the foregoing with a concrete example. Selznick explains how Communists operating in countries outside the socialist bloc would infiltrate non-Communist organizations, find their way into key positions within such organizations, and use those positions to influence the activity of the organizations in question. In some cases the organizations were taken over completely and made into appendages of the Communist Party. The Communists did not find it necessary to place large numbers of their people in the organizations that they sought to influence or control; a relatively small number of individuals, strategically placed and well organized, could exercise great power.[177]

For an anti-tech movement today there can be no question of simply copying Communist tactics. But careful study of a book like Selznick's can lead to ideas such as the following ones:

An anti-tech organization will have some degree of affinity with radical environmentalism. Many people tend to associate the term "radical environmentalist" only with illegal groups like Earth Liberation Front (ELF), but here we apply the term to any individual or group advocating environmental solutions that are too radical to have any chance of acceptance by the mainstream in modern society. For example, Bill McKibben—author of *Enough: Staying Human in an Engineered Age*—is a radical environmentalist by our definition, though as far as we know his work has always been entirely legal. Since we've already emphasized that a revolutionary organization committed to open, political action should maintain strict legality (section 27), it follows that the members of such an organization should avoid any involvement in illegal actions by radical environmentalists. But this need not prevent anti-tech revolutionaries from participating in the legal activities of radical environmentalist groups and seeking positions of power and influence within such groups. This power and influence could be used to the advantage of an anti-tech organization in various ways. For example:

(i) The anti-tech organization may be able to find suitable recruits for itself among the members of radical environmentalist groups.

(ii) If a member of the anti-tech organization can find a place on the editorial board of a radical environmentalist periodical (for instance, the *Earth First! Journal*), he will be able to influence the content of the periodical. If a majority of anti-tech people can be placed on the editorial board, they will be able in effect to take the periodical over, minimize its leftist content, and use it systematically for the propagation of anti-tech ideas.

(iii) If an anti-tech organization decides to undertake action on an environmental issue as suggested in section 17 of this chapter, and if it has power and influence within radical environmentalist groups, then it should be able to secure support and cooperation from these groups in carrying out the action in question.

(iv) In some cases the anti-tech revolutionaries may be able to take over a radical environmentalist group altogether and turn it into an anti-tech group. Under these circumstances leftists can be expected to drift away from the group, and in their place the group will attract recruits who are predisposed to anti-tech.

(v) Work in radical environmentalist groups will provide anti-tech revolutionaries with valuable training and experience in leadership and organizational work.[178]

(vi) When an acute crisis of the system arrives, the power and influence that anti-tech revolutionaries wield within radical environmentalist groups will be useful in the effort to organize on a mass basis.

None of this is inconsistent with the rule that the anti-tech movement must maintain clear lines of demarcation between itself and other radical movements. Lenin's emphasis on such lines of demarcation did not prevent him from collaborating—when he found it useful—with leaders of groups whose programs were in conflict with that of his own group.[179] Of course, members of the anti-tech organization who are asked to work within radical environmentalist groups will have to be clearly aware of the importance of the lines of demarcation. They will need to understand that their purpose in working with radical environmentalists is solely to win advantages for anti-tech and not to promote any radical environmentalist goals that may be inconsistent with anti-tech goals.

How can anti-tech revolutionaries get themselves into positions of power and influence in radical environmentalist groups? The most important way will be through

> the moral authority of hard work. In every organization which they seek to capture, the communists are the readiest volunteers, the most devoted committee workers, the most alert and active participants. In many groups, this is in itself sufficient to gain the leadership; it is almost always enough to justify candidacy [for leadership].[180]

> The [Communists] in penetrating an organization... become the 'best workers' for whatever goals the organization seeks to attain.[181]

This approach can be supplemented with a technique that Nelson Mandela used with outstanding success to get and keep leadership of the anti-apartheid movement in South Africa: He strictly controlled his emotions, rarely allowed himself to show anger, remained always calm, self-possessed, even-tempered.[182] This kind of deportment wins respect and encourages others to look to an individual for leadership. Among the Andaman Islanders, a potential chief was "a young adult in the camp who possessed the virtues that attract even younger men to seek his company. He was usually a good hunter, generous, *and, above all, even-tempered.*"[183]

A revolutionary working in a radical environmentalist group won't need to conceal his anti-tech commitment. But for obvious reasons he must avoid pushing anti-tech ideas aggressively, and he must not show disrespect for radical environmentalists' ideas. If he argues in favor of anti-tech he must do so in a good-humored way, and if an ideological discussion becomes heated or angry he must withdraw from it.

For the present this writer is not actually recommending that an anti-tech organization should use these methods to gain power and influence within the radical environmentalist movement. The leaders of an anti-tech organization will make that decision when the time comes, and they will take into account the resources of their organization, the opportunities available to it, and any other relevant factors. The point here is simply that the ideas outlined in this section are at least worthy of serious consideration, and that this writer would never have thought of those ideas if he hadn't studied Selznick's book. This example shows how the histories and the techniques of past movements can be an important source of ideas for an anti-tech movement today.

30. A revolutionary organization will need a section or a committee devoted to studying technology and keeping up with technological developments, and not only for the purpose of attacking technology politically. The organization also needs to be able to apply technology for its own revolutionary purposes.

It is well known that in the United States (and probably in most other countries) law-enforcement and intelligence agencies have long made use of wire-tapping—often illegally—to keep track of the plans and activities of politically suspect groups. But nowadays old-fashioned tapping of telephone lines is becoming obsolete and far more sophisticated eavesdropping techniques are available,[184] along with such tools for spying as ubiquitous surveillance cameras, face-recognition technology, hummingbird-sized (perhaps even insect-sized) drones, and mind-reading machines.[185]

In the United States, eavesdropping or spying by a government agency, unless authorized by a court of law, violates the Fourth Amendment's prohibition of unreasonable searches, and at least in some cases is illegal. But in all of the extensive legal research that this writer has conducted in relation to constitutional rights, he has never come across a single case in which government agents have actually been prosecuted for illegal eavesdropping or spying. While a civil lawsuit might theoretically be possible in some cases, we can say for practical purposes that almost the only legal defense against the government's illicit surveillance consists in the fact that evidence obtained in violation of the Fourth Amendment cannot be used in a criminal prosecution against the victim of the violation.[186] But there will be no prospect of criminal prosecution of members of a revolutionary organization that carefully maintains the legality of its activities. Consequently, government agencies will have no incentive to refrain from eavesdropping or spying on such an organization in disregard of the Fourth Amendment. Unconstitutionally and illegally acquired knowledge of the plans and activities of the organization may give the authorities a decisive advantage and enable them to sabotage the organization's efforts in various legal or illegal ways (as was done, for example, in the COINTELPRO program that we mentioned in section 27). Revolutionaries therefore need to be well informed about eavesdropping and spying technology, and need to have the technical capacity to defend themselves against its illegal use.

As time passes, it becomes less and less likely that revolutions in technologically advanced countries can be consummated by traditional methods; for example, by crowds of people taking to the streets. A careful study has shown that, for the traditional type of revolution, aid to the revolutionaries by elements of the military, or at least the neutrality of the latter, is usually required for success.[187] In the "Arab Spring" revolution of 2011 in Egypt, for instance, it is probable that the top military leaders gave in to many of the protesters' demands only because they feared that if it ever came to a showdown and they found it necessary to order crowds to be machine-gunned,[188] many of their troops would refuse to obey and might even defect to the revolutionaries. But techniques of crowd control are becoming ever more sophisticated: People can now be dispersed or incapacitated with superpowerful sound-blasters and strobe torches,[189] and a soldier who would refuse to shoot into a crowd of his fellow citizens might have no qualms about blasting them off the streets with unendurable volumes of sound. Following a riot, police will be able to track down participants with the help of images from surveillance cameras, face-recognition technology, and records of telephone traffic.[190]

More importantly, the replacement of humans by machines in the military is proceeding apace.[191] At the moment, human soldiers and policemen are still necessary, but, given the accelerating rate of technological development, it is all too possible that within a couple of decades police and military forces may consist largely of robots. These presumably will be immune to subversion and will have no inhibitions about shooting down protesters.

Of course, technology can be used by rebels, too, against the established power-structure.[192] Thus, a future revolution probably will not be carried out in the same way as any of the revolutions of the past or present. Instead, the outcome will depend heavily on technological manipulations, both by the authorities and by the revolutionaries. The importance for revolutionaries of technological competence is therefore evident.

NOTES

1. Selznick, p. 113.
2. See Kaczynski, Letters to David Skrbina: Aug. 29, 2004, last three paragraphs; Sept. 18, 2004; March 17, 2005, Part II.B.
3. See Hoffer, §§ 104, 110, 111.
4. Smelser, p. 353, notes that when the existing social order loses its appearance of invulnerability, new possibilities for revolution may open up.
5. See Chapter One.

6. Horowitz, p. 126. Horowitz, pp. 63–65, describes how Fidel Castro groped his way through the Cuban Revolution, learning from experience as he went along.

7. Matthews, p. 123. I seriously doubt that Hitler could have outlined his program with any degree of precision. "In 1928, before the onset of the Great Depression in Germany, Hitler received less than 3 percent of the vote," and it was the Depression that enabled him to become powerful. NEB (2003), Vol. 27, "Socio-Economic Doctrines and Reform Movements," p. 416. I've had neither the opportunity nor much inclination to read *Mein Kampf*, but I find it hard to believe that Hitler, in the early 1920s when he wrote his book, could have predicted the occurrence and the approximate time of the Depression.

8. See Alinsky, pp. 5–6, 45, 69, 136, 153–55, 164, 165–66, 168, 183.

9. Trotsky, Vol. Three, Appendix Two, p. 409.

10. Radzinsky, p. 202. NEB (2010), Vol. 22, "Lenin," p. 934.

11. Trotsky, beginning on p. 298 of Vol. One and all through the rest of the history of the Revolution.

12. Hoffer, § 89. Smelser, p. 381. Trotsky, Vol. One, p. xviii.

13. Selznick, p. 23n6 (quoting Lenin, "A Letter to a Comrade on our Problems of Organization," in *Lenin on Organization*, pp. 124–25). Alinsky, pp. 77–78, 120.

14. NEB (2003), Vol. 28, "Union of Soviet Socialist Republics," p. 1000.

15. Trotsky, Vol. One, Chapt. VIII, pp. 136–152. Cf. Kaczynski, Letter to David Skrbina: Sept. 18, 2004, fourth paragraph.

16. NEB (2003), Vol. 29, "War, Theory and Conduct of," pp. 647, 660. Mao, pp. 58–61. Dunnigan & Nofi, p. 54. Parker, p. 316.

17. Dorpalen, p. 332.

18. Trotsky, Vol. Two, p. 315. See also Selznick, pp. 22, 70, 217.

19. Mao, pp. 78–79.

20. See Chapter Three, Part IV.

21. Radzinsky, p. 82.

22. NEB (2010), Vol. 22, "Lenin," p. 936. NEB (2003), Vol. 28, "Union of Soviet Socialist Republics," p. 1002. Radzinsky, p. 375.

23. NEB (2010), Vol. 3, "Corday, Charlotte," p. 624.

24. Kee, pp. 564, 578.

25. Kaufmann, editor's preface to "Thus Spoke Zarathustra," p. 111.

26. Ulam, pp. 551–52. Thurston, e.g., pp. 163, 215, 225–26, 282n76. NEB (2003), Vol. 29, "World Wars," pp. 1009, 1023 (table).

27. Keegan, pp. 420–432. *World Book Encyclopedia* (2011), Vol. 21, "World War II," p. 482 ("German morale failed to crack"). But: Parker, p. 345 (strategic bombing "had a significant impact on German morale"). Gilbert, *European Powers*, pp. 264, 266. Manchester, pp. 527–29 ("Winston Churchill promised the

Commons that Germany 'will be subjected to an ordeal the like of which has never been experienced by a country in continuity, severity, and magnitude.'"), 647–48. NEB (2003), Vol. 29, "World Wars," pp. 1020, 1024. See also Paz. In contrast to that of German civilians, the morale of the Japanese civilian population was "brought to breaking-point" by a bombing campaign similar to that carried out over Germany. Keegan, p. 432. The reason was probably that Japanese housing (unlike German) was built of wood, with the result that incendiary bombing had a far more devastating effect on Japanese cities than on German ones. Dunnigan & Nofi, p. 109.

28. Wolk, p. 5. This may be an extreme example, but the usual attrition rate among Allied air crews was severe enough. Keegan, p. 433. Astor, p. 360. During 1943, "some eighty-three per cent [of British bomber crews] were failing to complete unscathed their tours of thirty operations. Of courage they had plenty, but there was nothing but lip-biting gloom registered on those faces." A. Read & D. Fisher, p. 127. During summer & autumn 1943, the Americans lost 30% of their bomber crews every month. Parker, p. 345. See also *World Book Encyclopedia* (2011), loc. cit.; Parker, p. 346. It's true of course that air crews' morale did suffer when attrition became excessive. Keegan, p. 428. Astor, loc. cit.

29. Murphy, passim. Dunnigan & Nofi, pp. 403, 625–26. It's worth noting that—apart from the acute risk of being killed or crippled—the physical hardships suffered by American soldiers in World War II were minor in comparison with what other soldiers in other wars have suffered. E.g., when Washington's defeated, starving, and half-naked army went into winter quarters at Valley Forge in 1777, many of the men had no shoes, so that "the soldiers of the Revolution [could be] tracked by the blood of their feet on the frozen ground." Martin, pp. 58, 161. The accuracy of Martin's memories, written down half a century after the events, may well be questioned, but sober history confirms that on the way to Valley Forge thousands of Washington's men were "barefoot and otherwise naked" and that the following winter was one of "semi-starvation." NEB (2003), Vol. 29, "Washington, George," p. 703. Yet the core of the rag-tag army held together and lived to fight again.

30. Murphy, pp. 4–8.

31. Kee, p. 732.

32. Stafford, p. 193.

33. Shattuck, p. 21.

34. De Gaulle, pp. 461–62.

35. *Polish American Journal*, Sept. 2012, p. 8; Feb. 2013, pp. 4, 7. Knab, pp. 1, 6. Lukowski & Zawadzki, pp. 261–63. For balance, it should be noted that there was a great deal of anti-Semitism in Poland at the time, and some Poles even helped the Nazis to round up Jews. Thurston, p. 224. Lukowski & Zawadzki, loc. cit. Interestingly, one of the two women who founded Zegota, an underground

organization dedicated to saving Jews, had been "generally considered... anti-Semitic" before the war. Jacobson, p. 7.

36. Woo, p. 11B.

37. P.R. Reid, p. 11.

38. De Gaulle, p. 713.

39. Hammer, p. 69.

40. Lee Lockwood, p. 80.

41. Gallagher, p. 45.

42. Bolívar, "Memoria dirigida a los ciudadanos de la Nueva Granada por un Caraqueño," in Soriano, p. 54.

43. Alinsky, p. xxii. See also p. 189 (referring to "a willingness to abstain from hard opposition as changes take place").

44. See Kee, pp. 519–592.

45. NEB (2003), Vol. 27, "Socio-Economic Doctrines and Reform Movements," p. 416.

46. See note 128 to Chapter Two.

47. See, e.g.: *The Economist*, July 16, 2011, p. 59; Sept. 10, 2011, p. 77. *The Week*, April 13, 2012, p. 16. *USA Today*, Sept. 27, 2012, p. 6B.

48. Smelser, p. 381.

49. Marx & Engels, pp. 21–22 (Introduction by Francis B. Randall), 46 (Engels's preface to English edition of 1888). Dorpalen, p. 211.

50. Kee, pp. 391, 405, 440–564. E.g., p. 537 ("Redmond... continued rightly to advise Birrell that these extremist forces represented only a minute proportion of Irish opinion...").

51. I've been told that Castro is thus quoted by Pandita, p. 35. I have not seen Pandita's book, but the quote is confirmed, to a close approximation, by Shapiro, p. 139, and the original source is given as "*N.Y. Times*, 22 Apr. 1959."

52. Estimates of the number range from seven to fifteen. Horowitz, p. 26. Russell, pp. 22, 23, 116, 117. NEB (2003), Vol. 2, "Castro, Fidel," p. 941. See also Matthews, pp. 93–98.

53. For Batista's army, NEB (2003), loc. cit., gives the figure 30,000. Russell, pp. 17, 22–23, cites estimates ranging from 29,000 to 50,000 (plus 7,000 police). For the size of Castro's force, NEB (2003), loc. cit., says 800 men. Estimates cited by Russell, pp. 23, 163, confirm that until just before Batista's fall the maximum size attained by the force under Castro's own command was about 800, but indicate that there were other guerrilla bands not under Castro's direct control, so that the total number of guerrilleros was somewhere between 1,000 and 1,500. At the very end of the rebellion "thousands" (the figures 8,000 and 40,000 are mentioned) "joined" Castro (though no reason is given to believe that most of these were under Castro's own control). Russell, pp. 23, 116, 163. But this did not happen until *at most* a few days before Batista fled the country on Jan. 1,

1959. Ibid. In other words, it was only after Batista had already been effectively defeated that "thousands" jumped on the revolutionary bandwagon.

54. For this whole paragraph see the following: NEB (2003), loc. cit. ("Castro's propaganda efforts proved particularly effective..."). Horowitz, pp. 62–65, 71–72, 127, 181 ("Castro's ability to manipulate the media is famous."). NEB (2003), Vol. 21, "International Relations," p. 865, says: "Fidel Castro took to the Sierra Maestra... and made pretensions of fighting a guerrilla war. In fact, Castro's campaign was largely propaganda..., and the real struggle for Cuba was fought out in the arenas of Cuban and American public opinion." According to Carrillo, p. 65: "[T]he victory of the 26th of July Movement... was possible because that movement was not a socialist party but a kind of national front that later split as the movement advanced, and in which the powerful personality of Fidel Castro and his closest collaborators brought about a subsequent turn toward socialism, while the right-wing sector openly went over to the American side." Information in greater detail is provided by Russell, pp. 17–28, 40–41, 78, 88, 115–120, 162–64.

55. Matthews, p. 96.

56. Gilbert, *European Powers*, p. 24.

57. See ibid.; NEB (2010), Vol. 22, "Lenin," pp. 933–34; Selznick, p. 176n2; and the quotation of Lenin cited in note 58 below. In 1894, according to Lenin, "you could count the [Russian] Social-Democrats on your fingers." This in "What Is to Be Done?," Chapt. III, section E; in Christman, p. 118. But no Social Democratic Party formally existed in Russia until 1898 at the earliest. Ulam, pp. 33, 49. Stalin, *History of the Communist Party*, first chapter, Section 4, p. 34. But see also ibid., Section 3, p. 29.

58. Selznick, pp. 103–04, quoting Lenin, "Lecture on the 1905 revolution," in *Collected Works*, 1942 edition, Vol. 19, pp. 389–390.

59. Trotsky, Vol. One, pp. 37, 40.

60. Radzinsky, pp. 133–34, 234.

61. The "one-sixth" figure is often cited; e.g., by Trotsky, Vol. Two, p. 121; Stalin, *History of the Communist Party*, Conclusion, p. 484; and Ulam, p. 288. But these writers fail to note that the figure is correct only if Antarctica is excluded.

62. Trotsky, Vol. Two, p. 306. Compare Ulam, p. 140 (the Bolsheviks were "a determined and disciplined party... at least in comparison with others..."), p. 143 (referring to "the bumbling behavior of the leaders of the... Mensheviks and Socialist Revolutionaries..."), p. 155 (referring to "the Mensheviks' and Socialist Revolutionaries' indecision and paralysis of will").

63. Trotsky, Vol. One, p. 398.

64. This refers to August and September 1917. Ibid., Vol. Two, p. 282.

65. See NEB (2010), Vol. 22, "Lenin," p. 936. Trotsky, Vol. Three, pp. 76, 88–123, 294, gives the impression that by October 1917 the Bolsheviks had won the support of the great majority of the Russian population, or at least of the

peasants, the soldiers, and the proletariat. But Trotsky probably felt compelled for ideological reasons to portray the Bolsheviks as having the support of "the people." It's more likely that, even at the best of times, the Bolsheviks had the active support only of some smallish minority and the mere passive acquiescence of a larger number (possibly though not necessarily a majority), while most of those who feared or disliked the Bolsheviks were disorganized and intimidated, therefore ineffective.

66. Actually November according to modern dating. Prior to the Revolution Russia used "Old Style" dates, i.e., dates according to the Julian Calendar, while most of the rest of the world was using the Gregorian Calendar, the calendar that is still in use today. In this book we haven't bothered to distinguish between Old Style and New Style dates in Russian history, because the difference of some 13 days is of no importance for our purposes. Readers who want accurate dates can refer to any history of the Russian Revolution.

67. NEB (2003), Vol. 28, "Union of Soviet Socialist Republics," p. 1003.

68. Trotsky, Vol. Three, p. 294.

69. Ulam, pp. 178–79.

70. Trotsky, Vol. One, pp. 137–140; Vol. Two, p. 302 ("The Petrograd [St. Petersburg] Soviet [was] the parent of all the other soviets. . . ."); Vol. Three generally, especially pp. 88–123. Ulam, p. 137. Stalin, *History of the Communist Party*, seventh chapter, Section 6, p. 286.

71. See ISAIF, ¶ 196.

72. Trotsky, Vol. Three, pp. 173, 284.

73. Ibid., p. 130.

74. Alinsky, p. 107.

75. Trotsky, Vol. Two, pp. 4, 7.

76. Alinsky, pp. 77–78, 113–14, 120, 128–29.

77. Trotsky, Vol. One, pp. xviii–xix. See also ibid., p. 110 ("A revolutionary uprising... can develop victoriously only in case it ascends step by step, and scores one success after another. A pause in its growth is dangerous; a prolonged marking of time, fatal."). I can't pretend to say under just what circumstances these dicta of Trotsky's are actually valid, but there are numerous counterexamples to them unless the term "revolutionary uprising" is interpreted very narrowly. It remains true, however, that momentum is a very important factor in revolution, as it is in many other conflict situations.

78. Trotsky, Vol. Two, pp. 9–31, 63, 68, 73, 82–83. Ulam, pp. 144–48, portrays the Bolsheviks' action during the July Days as much more confused and less calculated.

79. Alinsky, pp. 27–28, 78, 133–34. See also ISAIF, ¶ 186.

80. "In a serious struggle there is no worse cruelty than to be magnanimous at an inopportune time." Trotsky, Vol. Three, p. 215.

81. Ibid., Vol. One, pp. 323–24.

82. Ibid., Vol. Two, p. 453.

83. Ibid., p. 306.

84. Ibid., Vol. Three, p. 166.

85. See note 26 to Chapter One.

86. Nissani, Chapt. 2, especially pp. 62–69. NEB (2003), Vol. 8, "nuclear winter," p. 821. Shukman, pp. 44–45. G. Johnson, pp. 126, 128–29.

87. See Rule (iii) of Chapter Three; Alinsky, p. 113 ("power comes from organization.... Power and organization are one and the same.").

88. This is essentially what the dispute between Lenin and Martov was about. Selznick, p. 57&n43. Stalin, *History of the Communist Party*, second chapter, Section 3, pp. 60–61. Ulam, p. 52&n5, maintains that Lenin's dispute with Martov over control of the party journal, *Iskra*, was far more important, but what matters for our purposes is that Lenin proved to be right about the composition of the party.

89. See the discussion of Rule (iv) in Chapter Three, Part III.

90. Cf. Smelser, pp. 120–22, 313–325.

91. Trotsky, Vol. Three, p. 166.

92. See Trotsky, Vol. Two, p. 311 ("strength is accumulated in struggle and not in passive evasion of it").

93. Mao, p. 161. Hitler would have agreed. See Hoffer, § 73.

94. Selznick, p. 132, quoting from a Communist document.

95. Selznick, p. 49, quoting Lenin, "Where to Begin," in *Collected Works*, 1929 edition, Vol. 4, Book I, p. 114.

96. Fidel Castro, letter of Aug. 14, 1954, in Conte Agüero; quoted in Horowitz, pp. 62–63.

97. Stalin, *Foundations of Leninism*, pp. 116–17, quoted by Selznick, p. 35.

98. Sampson, p. 427.

99. Ibid., p. 50. See also pp. 403, 427 (Mandela always regarded himself as a "loyal and disciplined member" of the ANC).

100. In Chapter One, Part III, we've called attention to the fact that the formally empowered leaders of an entire nation have in reality only limited power over the functioning of their society, and of course the leaders of any organization face a similar problem to a greater or lesser degree. So the question arises of the extent to which the leaders of an anti-tech organization will actually be able to control it. I won't attempt a serious discussion of this difficult subject, but will merely point out that the problem of control is far less acute in the case of our revolutionary organization than it is in the case of an entire nation or even of an entity such as a large corporation. To mention only one reason, the revolutionary organization will be to a great extent ideologically uniform because its members are to be recruited selectively (see section 14, above), troublesome members will

be relatively easy to identify and expel, and any dissident faction that may develop should withdraw to form a separate organization (see section 19). This will tend strongly to reduce the conflict of individual wills within the organization.

Later, when the revolutionaries assume leadership of a mass movement, the problem of control may indeed be acute. (Recall for example the case of the "July Days"—mentioned above, section 12—in which the Bolsheviks were able to prevent an untimely insurrection only at very great cost to themselves. See Trotsky, Vol. Two, pp. 1–84, 250–58.) On the other hand, even at this stage, the fact of being locked in a hard struggle against external adversaries will tend to unite the movement behind its leaders, and this will facilitate control.

In well-organized revolutionary movements such as those of the Bolsheviks and the Nazis, the core of the movement (though not necessarily the mass following) seems to have remained, generally speaking, well under the control of the leaders prior to the time when the movement came into power. But once the movement had assumed the government of an entire nation, grave problems of control did emerge. See Chapter One, Part III; Chapter Three, passim.

101. "[N]o revolutionary organization has ever practiced *broad* democracy, nor could it, however much it desired to do so." Lenin, "What Is to Be Done?," Chapt. IV, Part E; in Christman, p. 161; and see the example of the British trade unions on the succeeding pages.

102. See Selznick, pp. 96–97&n17, 288n15.

103. See Schebesta, II. Band, I. Teil, p. 8; Turnbull, *Forest People*, pp. 110, 125, and *Wayward Servants*, pp. 27, 28, 42, 178–181, 183, 187, 228, 256, 274, 294, 300.

104. Trotsky, Vol. Two, p. 306.

105. Ibid., Vol. One, pp. 306–313.

106. NEB (2010), Vol. 22, "Lenin," pp. 933–34. Though the name "Menshevik" (from "menshe" = "smaller") implies minority status and "Bolshevik" (from "bolshe" = "bigger") implies that the Bolsheviks were a majority, the Bolsheviks were in fact a minority. Ibid., p. 933. Christman, editor's introduction, p. 6. See also Ulam, p. 50, and Lenin, "The State and Revolution," Chapt. IV, section 6; in Christman, p. 332.

107. NEB (2010), Vol. 22, "Lenin," p. 934. Here it is stated that "not a few Bolsheviks supported the war effort." Hence, the "closest comrades" who followed Lenin on this issue at first comprised only some subset of the Bolsheviks. See Ulam, pp. 126–28.

108. NEB (2010), Vol. 22, "Lenin," p. 935.

109. Trotsky, Vol. One, pp. 298–312.

110. Ibid., Vol. Three, pp. 124–166.

111. Alinsky, pp. 60, 79.

112. Ibid., pp. 19, 105–06, 113–14, 117–19, 178, 194.

113. Trotsky, Vol. Three, p. 73.

114. Selznick, p. 39 (quoting Dimitrov, p. 124). After the Revolution the Bolsheviks changed their name to "Russian Communist Party (Bolsheviks)." Selznick, p. 10n3, says the name was changed in 1919, but this is an error. Ulam, p. 168&n8, Stalin, *History of the Communist Party*, seventh chapter, Section 7, p. 299, and other sources agree that the name was changed in 1918. Later there were further changes of name, Ulam, pp. 703–04, 732, but all of the new names contained the phrase "Communist Party." Accordingly, we here use the term "Communist" to refer to the post-1918 Bolsheviks, their adherents, and their successors.

115. Cf. ISAIF, ¶ 141.

116. Trotsky, Vol. One, p. 294; Vol. Two, pp. 310–11; Vol. Three, p. 127, Appendix One, p. 376.

117. Ibid., Vol. Two, p. 312.

118. Ibid., Vol. Two, p. 320; Vol. Three, pp. 127–28.

119. Mao, p. 346 (editors' note at the foot of the page).

120. Mao, p. 397. See also p. 189.

121. Trotsky, Vol. Two, p. 251. See also Alinsky, p. 114.

122. This cursory account has been pieced together from two sources that are not perfectly consistent with one another: Kendrick, pp. 237–39 and MacFadyen, Chapts. IV, V.

123. NEB (2003), Vol. 1, "Alfred," p. 260.

124. For the whole story see Barrow, Duncan, and NEB (2003), Vol. 29, "United Kingdom," pp. 40–41, 120. John Barbour's poem is by no means accurate historically, but the editor, Duncan, provides copious notes in which he tries to sort out fact from legend.

125. Barrow, pp. 154, 160–61, 164, 166–171. Barbour, Books 2, 3, in Duncan.

126. Barrow, p. 166.

127. Ibid., p. 187.

128. Ibid, p. 165.

129. Lidtke, pp. 77–81.

130. Ibid., p. 81.

131. Ibid., pp. 89–97.

132. Ibid., pp. 97–104.

133. Ibid., p. 185. Compare the figures on this page with those on p. 74.

134. Mao, p. 307.

135. Ibid., p. 309n6 and pp. 177–78n3.

136. Sampson, p. 259.

137. See Selznick, p. 10n3, pp. 103–04; NEB (2010), Vol. 22, "Lenin" pp. 933–34.

138. Gilbert, *European Powers*, p. 25.

139. Radzinsky, p. 90 (quoting an old witness who had lived through the events).

140. Gilbert, loc. cit.

141. Trotsky, Vol. One, pp. 36–37.

142. Ibid., Vol. Two, pp. 1–84.

143. Ibid., pp. 250–58.

144. Ibid., p. 256.

145. Ibid., p. 258.

146. Radzinsky, p. 324.

147. Matthews, p. 95.

148. Ibid., p. 31.

149. Ibid., pp. 96–97.

150. E.g., Mao, pp. 476–78; Saney, pp. 19–20; Christman, editor's introduction, p. 4; Ulam, p. 293.

151. Duncan, p. 120 (editor's note to Barbour's Book 3).

152. That personal ambition was Bruce's principal motive can be inferred from Barrow, pp. xii, 17–18, 33, 41–44, 84, 110, 121–22, 124, 141, 142–44, 146, 150, 174, 200, 202, 245, 254, 262, 313.

153. Cruelties inflicted on English: Barrow, pp. 197, 236, 240, 243, 248, 254, 256, 262; on Scots: pp. 174, 175–77, 181–82, 189, 190, 194, 256; on Irish, p. 315. See also Duncan, loc. cit. and passim.

154. See Kee, pp. 351–470.

155. See NEB (2003), Vol. 26, "Propaganda," pp. 176, 177.

156. Ibid., p. 176.

157. "[A] great deal of [Lenin's] writing is devoted to the drawing of lines between his group and others... there was this great emphasis on sharp differentiation... ." Selznick, p. 127. See Lenin, "What Is to Be Done?," Chapt. I, section D; in Christman, pp. 69–70.

158. See ISAIF, ¶¶ 6–32, 213–230; Kaczynski, Letter to M.K.; *Green Anarchy* # 8, "Place the Blame Where It Belongs," p. 19; Kevin Tucker's letter to the editors of *Anarchy: A Journal of Desire Armed* # 62 (Fall-Winter 2006), pp. 72–73; and this writer's own letter to the editors of the same journal, # 63 (Spring-Summer 2007), pp. 81–82.

159. Kaczynski, Preface to the First and Second Editions, points 3 & 4, and ISAIF, ¶¶ 213–230.

160. Matthews, p. 103.

161. "COINTELPRO" stands for "Counterintelligence Program." For information about COINTELPRO, see *Select Committee to Study Governmental Operations With Respect to Intelligence Activities*, *Final Report*, S. Rep. No. 755, Book II (Intelligence Activities and the Rights of Americans) and Book III (Supplementary Detailed Staff Reports on Intelligence Activities and the Rights

of Americans), 94th Congress, Second Session (1976). Also, *Hobson v. Wilson*, 737 F2d 1 (D.C. Cir. 1984) (this means Vol. 737, *Federal Reporter*, Second Series, p. 1, United States Court of Appeals for the District of Columbia Circuit, 1984).

162. *Warrior Wind* No. 2, pp. 1–2 (available at the University of Michigan's Special Collections Library in Ann Arbor). The informer may even have been an agente provocatrice. Ibid.

163. Lipsher, pp. 1A, 25A. Three of the names mentioned in Lipsher's article (Gerlach, Ferguson, Rodgers) are also mentioned in *Warrior Wind* No. 2, pp. 5, 8.

164. Sampson, pp. 170, 171, 183, 245–47, 254, 258–260, 313–14, 387.

165. Kee, pp. 563, 648.

166. Matthews, pp. 102–03.

167. Guevara, e.g., p. 261.

168. Lidtke, p. 94.

169. Ibid., p. 93.

170. Ibid., p. 98.

171. Ulam, e.g., pp. 87, 95n16, 107, 114. Interesting information about the methods of the Tsar's secret police can be found in Vassilyev.

172. Pipes, p. 25n2. Ulam, pp. 72–73.

173. Trotsky, Vol. One, p. 37.

174. Ulam, pp. 113, 114, 121, 123, 125&n19, 320. Pipes, pp. 24–25. It may be, however, that Lenin "allowed for" the possibility that Malinovsky was a spy, "but thought that... the Bolsheviks benefited more than the police from his duplicity." Ibid.

175. Kaczynski, Letters to David Skrbina: Sept. 18, 2004, second paragraph; Jan. 3, 2005, Fifth point.

176. Mao, pp. 58–59, 61–62, 71–72, 77–80, 198–208. Not everything in this paragraph was explicitly stated by Mao, but all of it can be inferred from what he did state explicitly.

177. All this is a major theme of Selznick's book. See, e.g., pp. 66–67, 90, 118–19, 150–54, 171–72, 175, 189–190, 208–09, 212&n43.

178. Cf. ibid., p. 19.

179. Ibid., pp. 126–28.

180. Ibid., p. 250.

181. Ibid., p. 319. Of course, anti-tech people in a radical environmentalist group will be able to work only for those radical environmentalist goals that do not conflict with the goals of their own anti-tech organization.

182. Sampson, pp. 210, 215, 242, 337, 491, 574. Azorín in his recommendations for political leadership, Sections XIV, XLV, emphasized these same qualities of equanimity and self-control.

183. Coon, p. 243 (emphasis added).

184. See, e.g., *The Week*, Oct. 8, 2010, p. 8, and April 13, 2012, p. 16.

185. E.g.: Cameras: *The Week*, Sept. 9, 2011, p. 14; *USA Today*, Jan. 4, 2013, p. 7A. Face recognition: *The Economist,* July 30, 2011, p. 56. Drones: *Time*, Oct. 22, 2007, p. 17 and Nov. 28, 2011, pp. 66–67; *The Week*, Jan. 14, 2011, p. 20, March 4, 2011, p. 22, Dec. 23, 2011, p. 14, June 15, 2012, p. 11, and June 28, 2013, pp. 36–37; *The Economist*, April 2, 2011, p. 65; *Wired*, July 2012, pp. 100–111; *Air & Space,* Dec. 2012/Jan. 2013, pp. 32–39; Ripley, pp. 67–74. Mind-reading machines: *The Economist*, Oct. 29, 2011, pp. 18, 93–94; *Time*, Nov. 28, 2011, p. 67; *The Week*, Feb. 17, 2012, p. 23; *USA Today*, April 23, 2014, p. 5A. Massive collection of data on individual citizens: *The Week*, Jan. 29, 2010, p. 14 and Sept. 17, 2010, p. 15; *USA Today*, Jan. 7, 2013, p. 6A. The facts revealed by Edward Snowden have been so widely publicized that it hardly seems necessary to cite any articles, but as an example we mention *USA Today*, June 17, 2013, pp. 1A–2A. Miscellaneous: *The Atlantic*, Nov. 2016, pp. 34–35. The foregoing is only a sample. Anyone who wants to take the trouble can easily dig up unlimited amounts of scary stuff about surveillance.

186. This is the "exclusionary rule." In practice, the federal courts generally enforce the exclusionary rule reluctantly and tend to invent exceptions to it.

187. Russell (the entire book).

188. E.g., in 1923, French troops occupying the Ruhr opened up with a machine gun on a crowd of protesters, killing 13. Gilbert, *European Powers,* pp. 110–11. But these were not their fellow Frenchmen.

189. "New riot-control technology: The sound and the fury," *The Economist*, Aug. 13, 2011, p. 56.

190. E.g., "The BlackBerry riots," *The Economist*, Aug. 13, 2011, p. 52.

191. Milstein, pp. 40–47. Whittle, pp. 28–33. Markoff, "Pentagon Offers Robotics Prize," p. B4. *The Economist*, April 2, 2011, p. 65. *National Geographic*, Aug. 2011, pp. 82–83. *Time*, Jan. 9, 2012, p. 30. Cf. Kaczynski, Letter to David Skrbina: March 17, 2005, Part III.D.

192. E.g.: Acohido, "Hactivist group," p. 1B. Acohido & Eisler, p. 5A. *The Week*, Feb. 18, 2011, p. 6. *The Economist*, March 19, 2011, pp. 89–90; Dec. 10, 2011, p. 34. *USA Today*, June 1, 2011, p. 2A; June 11, 2012, p. 1A; July 2, 2015, p. 3B; Nov. 13–15, 2015, p. 1A. Ripley, pp. 70, 72.

APPENDIX ONE

In Support of Chapter One

A. In answer to the arguments of Chapter One, true-believing technophiles like Ray Kurzweil and Kevin Kelly are likely to answer: "Technology will solve all those problems! Human beings will be transformed step by step into man-machine hybrids (cyborgs), or even into pure machines, that will be incomparably more intelligent than their human ancestors.[1] With their superior intelligence, these beings will be able to use the technological miracles of the future to guide the development of their society rationally." However, none of the arguments of Chapter One (with one exception, noted below) depend on the limitations of human intelligence or on any weaknesses peculiar to human beings, so there is every reason to think that the arguments will remain valid for a society derived from the present one through the piecemeal replacement of humans by machines in the manner envisioned by Kurzweil.

B. The technophiles won't be rash enough to claim that any future technological miracle will make it possible for a society to predict its own development over any substantial interval of time. But some, perhaps, will point to the fact that the modern mathematical theory of control now makes it possible—in *some* cases—to design mechanisms that will keep a complex system on a fixed course even if only the *short-term* "effect of any potential control action applied to the system" can be predicted (though the effect must be predictable "precisely... under all possible environmental circumstances").[2] But in the context of control theory a system is called "complex" if "the efforts of many persons and the use of special technical equipment (computers) are required to draw the whole picture,"[3] and examples of "complex" systems are "[t]he launch of a spaceship, the 24-hour operation of a power plant, oil refinery, or chemical factory, the control of air traffic near a large airport."[4] In the present discussion we are dealing with an entirely different level of complexity. Any one power plant, oil refinery, or chemical factory is extremely simple in comparison with an entire modern society.

Actually, a careful reading of what the *Britannica* says about control theory[5] will give scant encouragement to anyone who might like to believe that the theory would make possible the rational control of the development of an entire society. Among other reasons, control theory generally "is applicable to any concrete situation... [only when] that situation can be described, with high precision, by a [mathematical] model," and the applicability of the theory is limited by "the agreement between available models and the actual behavior of the system to be controlled."[6] For control of an entire society one would need a precise mathematical model of, among other things, human behavior (or of the behavior of cyborgs or of machines descended from humans, which in Kurzweil's vision would be far more complex even than human beings themselves). In special contexts, as when one needs only statistical information about human behavior, adequate models may be possible. E.g., in a marketing study one may be unconcerned with the actions of individuals; one may need only such information as the percentage of consumers who will buy a given product in specified circumstances. But for control of an entire society one would need a precise mathematical model of the behavior of each single one of numerous persons (whether human persons or machine-persons) including, at the least, all those who occupy positions of special importance (political leaders, top-level government officials, military officers, corporation executives, etc.) and whose individual behavior interacts continuously with the society as a whole and has a significant effect on it.

All the same, let's make the extremely daring assumption that a precise mathematical model of our entire society could actually be constructed. Even so, it is wildly improbable that sufficient computing power could ever be available to handle the trillions upon trillions upon trillions of simultaneous equations that would be involved. Remember what we pointed out in Part II of Chapter One: that sixty trillion equations would be required just for prices in the U.S. economy alone, leaving out of account all other factors in U.S. and world society; and that even if some future society had enough computing power for control of the *present* society, it wouldn't have enough to control its *own* development, because the complexity of a society grows right along with its computing power. Finally, even if enough computing power were available, it would be impracticable to collect the stupendous amount of minutely detailed, highly precise information that would be required for insertion of the appropriate numbers into the equations.

Thus, it is safe to conclude that no society will ever be able to set up a mathematically designed control system that will keep the society forever on a fixed course of development. Let us nevertheless carry to the utmost extreme our generosity toward the true-believing technophiles: Let's grant them the impossible and assume that such a control system could successfully be designed. Even under this assumption we still run up against fundamental difficulties: Who is going to decide what objectives are to guide the design of the control system that is to keep the society on a fixed course of development, and what fixed course of development should be chosen for the society? And how will the society be induced to accept the control system and the chosen course of development? If the control system is to be approved by the public at large it will have to be a compromise solution that in trying to satisfy everyone will satisfy no one. In practice, it is unlikely that such a compromise could ever win general acceptance, so any control system would have to be forcibly imposed by an authoritarian faction that had acquired dictatorial power. In that case—let the citizen beware! Furthermore, if any faction ever became powerful enough to impose its own solution on society, it would probably be riven thereafter by internal power-struggles. (Recall the remark of Benjamin Franklin quoted in Part II of Chapter Two, and see Part C of Appendix Two, below.)

The notion of a future society governed in accord with a mathematical control system, rationally chosen and designed, can be dismissed as science fiction.

C. Let's take another look at the idea that we considered and disposed of in Part V of Chapter One: that of an all-powerful philosopher-king. In order even to entertain the notion that such a philosopher-king could rationally steer the development of a society, we already had to make assumptions that were wildly improbable. We then noted that, even granting those assumptions, we still ran into fundamental difficulties: that of selecting a satisfactory philosopher-king and putting him into a position of absolute power; and that of ensuring the succession, after the death of the original philosopher-king, of a long line of competent and conscientious philosopher-kings who would all govern in accord with some stable and permanent system of values.

The technophiles will have a ready answer to the second difficulty: They will argue that biotechnology will make it possible in the future to hold back the aging process indefinitely;[7] hence, our philosopher-king will be immortal and the question of choosing a successor will never arise. But this still doesn't solve the problem of rational guidance of a society's development, for people change over time, and our philosopher-king will change too. His decisions will affect the society in which he lives in ways that will not be fully predictable, and the changes in society will in turn affect the philosopher-king's goals and values in ways that cannot be foreseen. Consequently, the society's development over the long term will not be steered in accord with any stable system of values but will drift unpredictably.

At this point in our discussion—and only at this point—the distinction between human beings and intelligent machines becomes relevant: In place of a human philosopher-king, technophiles may propose rule by a super-computer hardwired to adhere forever to a fixed system of values. Even if we assume that such a computer could be created and that it would remain internally stable, we still face fundamental difficulties: Who is going to decide what values are to be hardwired into the electronic philosopher-king, and how will they put their electronic philosopher-king into a position of absolute power? This is no more easy to answer than the question discussed in Part V of Chapter One of how to choose a human philosopher-king and put him into a position of absolute power, or the question discussed in Part B of this appendix of how to choose a mathematical control system and secure the submission of society to its rule.

It would in any event be impossible to formulate a satisfactory system of values. Any values would be sure to give unsatisfactory results if they were sufficiently precise and rigid to determine the electronic philosopher-king's decisions in all cases without leaving the machine any substantial discretion to make its own value-judgments. This will be clear to anyone who has ever done much research in American constitutional law. The rules of decision laid down by the courts are full of vague "balancing tests" and indefinite "factors" on which judges are supposed to rely in deciding cases. Two judges applying the same "balancing test" or "factors" in the same case will often come to radically different conclusions; hence the numerous dissenting opinions that one finds in the published decisions of the U.S. Circuit Courts and the Supreme Court. The reason why the rules of decision are so vague and flexible is that it is impossible to formulate precise, rigid principles that will determine the outcome of all cases in even a remotely satisfactory way. If the courts were held strictly to any such set of rigid principles, they would be forced to make many decisions that practically everyone would regard as unreasonable.

On the other hand, if the system of values hardwired into the electronic philosopher-king were sufficiently vague or flexible to allow the machine any significant leeway to make its own value-judgments, gone would be the stability of values that the hardwiring was supposed to ensure. Where principles are in any substantial degree vague or flexible, one can usually find a way to justify almost anything in terms of them. Hence, two decisions that are both arguably in harmony with the same set of principles can have radically different practical consequences; this again is seen in the dissenting as against the majority opinions of the U.S. federal courts.

Thus, even apart from all other difficulties, the impossibility of formulating a satisfactory system of values is by itself sufficient to justify us in dismissing as science fiction the notion of a future society ruled by a supercomputer hardwired to govern according to a stable and permanent system of values, if the system of values is expected to give results that we would regard as even marginally acceptable.

D. The reader may wonder why we have even bothered with this excursion into science fiction. But for the problems facing our society today it is likely that technophiles will envision future solutions that to most people will look like science fiction. Ray Kurzweil's book, for example, is full of that type of material, and much of it is indeed science fiction. Nonetheless, it is always risky to dismiss ideas about future technological developments as science fiction solely because they seem implausible on vague intuitive grounds. Things that seemed implausible at the outset of the Industrial Revolution, or even just a few decades ago, are not the least bit implausible today. To mention only one example, back in the 1950s, when Moore's Law had never been heard of, most people, probably including most computer scientists, would have dismissed as implausible the suggestion that fifty years later every Tom, Dick, and Harry would hold comfortably in his lap more computing power than that of a whole roomful of 1950s computing machinery costing millions of dollars. Futuristic proposals need to be examined critically and dismissed as science fiction only when good reasons for the dismissal have been found.

But whatever technological miracles the future may have in store, we think there are excellent reasons for dismissing as science fiction the notion that the development of a society will ever be subject to rational guidance.

NOTES

1. Kurzweil, e.g., pp. 194–203, 307–311, 324–26, 374–77, 472.
2. NEB (2003), Vol. 25, "Optimization, Mathematical Theory of," p. 224.
3. Ibid., p. 223.
4. Ibid., p. 224.
5. Ibid., pp. 223–26.
6. Ibid., p. 224.
7. E.g., "Mr. Immortality," *The Week*, Nov. 16, 2007, pp. 52–53; Grossman, pp. 46–47; Kurzweil, pp. 9, 212–15, 371.

APPENDIX TWO

In Support of Chapter Two

A. Proposition 2 of Chapter Two states that in the short term, natural selection favors self-propagating systems that pursue their own short-term advantage with little or no regard for long-term consequences.

- Steven LeBlanc[1] argues that among primitive societies natural selection favors ecological recklessness. Suppose one group lives prudently within its resources while a neighboring group allows its population to grow to the point where its resources are over-strained, so that its environment is damaged and it can no longer feed itself adequately. In order to find an outlet for its surplus population, the second group may try to take the first group's territory by force, and it is likely to succeed, because it has more people and can put more warriors into the field than the first group can. "This smacks of a Darwinian competition—survival of the fittest—between societies. Note that the 'fittest' of our two groups was not the more ecological, it was the one that grew faster."[2] LeBlanc admits that his argument is oversimplified,[3] and certainly it is not applicable in all circumstances, but it does seem to contain a good deal of truth.
- During the 1920s the Soviets needed to acquire technological equipment from industrialized countries in order to catch up with the West economically, so they resorted to trade with Western capitalists.[4] One might have thought that capitalists would refuse to trade with communists, since the latter were bent on destroying capitalism, but in order to make a profit the capitalists were willing, as Lenin allegedly put it, to "sell the rope to their own hangmen."[5] In 1971, Alinsky claimed to "feel confident" that he could "persuade a millionaire on a Friday to subsidize a revolution for Saturday out of which he would make a huge profit on Sunday even though he was certain to be executed on Monday."[6] Alinsky was exaggerating for humorous effect, but his remark does reflect a truth about capitalism. It's easy to attribute the capitalists' shortsightedness to "greed," but there is a reason why capitalists are greedy: Those who forgo profit in the present from concern for long-term consequences tend to be eliminated by natural selection.

• The U.S. financial crisis that began in 2007 resulted from the widespread offering of risky ("subprime") loans to borrowers who needed the money to buy homes but might never be able to pay it back.[7] Lenders such as savings-and-loan associations sold the right to collect their subprime loans to other financial organizations, which sold the right in turn to still other organizations, and so forth, in a process much too complex to be described here. The subprime loan market was so lucrative and important that even the government-sponsored enterprise known as Fannie Mae feared "the danger that the market would pass [it] by"[8] if it refused to deal in subprime loans. Fannie Mae was so big and powerful that its survival would not have been threatened if it had not participated in the subprime loan market, but we can imagine that many smaller, private financial enterprises would have been unable to survive in the face of competition if they had failed to make use of the opportunities offered by subprime loans. However, for enterprises that did make use of those opportunities there was a terrible price to be paid when the housing bubble burst. Even the two gigantic government-sponsored enterprises, Fannie Mae and Freddie Mac, collapsed and had to be rescued by the government.[9] Needless to say, many private financial enterprises too went bankrupt.[10] What appears to have happened is that the pressure of competition forced these enterprises to take risks that later had fatal consequences. No doubt greed too was involved, but, as we pointed out a moment ago, capitalists who are not greedy tend to be eliminated by natural selection.

• In the modern world, international trade is highly important for the economic success of the nations involved;[11] it is even believed that no modern nation could survive economically if it did not participate in international trade.[12] But in the longer term such trade entails serious risks:

> [A] country that has become heavily involved in international trade has given hostages to fortune: a part of its industry has become dependent upon export markets for income and for employment. Any cutoff of these foreign markets... would be acutely serious; and yet it would be a situation largely beyond the power of the domestic government involved to alter. Similarly, another part of domestic industry may rely on an inflow of imported raw materials, such as oil for fuel and power. Any restriction of these imports could have the most serious consequences;[13]

and reliance on the importation of manufactured goods too can be risky.[14]

It's possible that Germany's dependence on international trade was a decisive factor in that country's defeat in World War I, for the British blockade was so effective in cutting off German trade that by the end of the war it had brought Germany to the verge of starvation.[15] On the other hand, Britain's dependence on international trade would have led to a German victory in either World War I or World War II if the British hadn't succeeded, with American help, in defeating Germany's submarine campaign, for the U-boats would otherwise have starved Britain into submission.[16] What we see, therefore, is that for the sake of economic survival in the short term nations must take the risk of allowing themselves to become dependent on international trade, even though their dependence may have grave or even fatal consequences in the long run.

- It is currently believed that the United States is "the most profligate or wasteful" of all developed countries in its use and abuse of its natural resources.[17] This has probably been true throughout U.S. history. In colonial times, American farming methods were recognized as highly improvident in comparison with European ones,[18] and Zimmermann points out the reckless and wasteful way in which, during the 1860s and 1870s, the fabled Comstock Lode in Nevada was exhausted within twenty years, whereas, says Zimmermann, a similar body of ore in Europe would have provided thousands of miners with a livelihood for centuries.[19] This was probably typical of American mining practices at the time. Yet America's profligacy in the use of its natural resources didn't prevent it from becoming the world's dominant economic power. And the country that is now beginning to challenge America's dominance is China, which is notorious for its environmental irresponsibility.[20] As these examples illustrate, reckless exploitation of natural resources can favor the achievement of power in the short term, however deadly its long-term consequences may be.

B. In connection with Propositions 4 and 5 of Part II of Chapter Two, we mentioned that pre-industrial empires spanning vast distances "actively created, if they did not already have, relatively rapid means of transportation and communication."

The Egyptians had the Nile. The Romans relied heavily on water transport over the Mediterranean and the rivers that flowed into it,[21] and for overland travel they built their famous roads. The Persians built a canal connecting the Mediterranean to the Red Sea, and a "Royal Road" that stretched 1,600 miles and made possible the quick delivery of letters by postal relays.[22] Imperial China, throughout its history, built and maintained canals, roads, and bridges and operated postal relays.[23] The Mongol empire of Chinggis (Genghis) Khan "utilized homing pigeons as messengers" and had "an extensive system of messenger posts" through which relays of riders carried messages at top speed.[24]

The Incas built roads and bridges over which relays of runners could carry messages rapidly, while freight was transported on the backs of human porters or llamas.[25] The Maya never created an empire of any substantial extent, and their lack of developed facilities for long-distance transportation or communication probably had something to do with this.[26] The Aztecs' system of long-distance communication was poorly developed: Messages were carried by relays of runners,[27] but there was no adequate system of roads for them, and some routes were probably impassable during the rainy season.[28] So it's not surprising that the Aztec "empire" (if it can be called that) was only weakly cohesive: Conquered peoples could be forced to pay taxes, or to contribute troops for Aztec campaigns and labor-gangs for Aztec work-projects, but in other respects there was very little centralized control.[29] Even at that, the "empire" appears to have reached the maximum geographical extent that was possible with the existing means of transportation and communication,[30] and it was probably unstable, for revolts were frequent.[31]

C. It seems clear in general that internal dissension within large human groups tends to be inversely proportional to the magnitude of external threats or challenges to the group, so that a dramatic reduction of external threats or challenges tends to be followed by a marked increase in internal dissension. "A social scientist, Michael Desch, ... noticed that external threats led to internal cohesion, and when the threat was removed, the cohesion broke down, sometimes violently."[32] This was hardly an original observation on Desch's part. But here, as so often elsewhere, "clean" historical examples are scarce, due to the complexity of historical developments in the real world. See note 7 to Chapter Two. However, we offer four relatively clean examples:

- "The general view of thinking Romans was that the relaxation of external pressures" due to "the temporary end of the age of major wars (ca 130 BC)" was what led to the "internal disintegration" of the Roman Republic.[33] Though the *Britannica* seems uncertain, it's hard to believe that the relaxation of external pressures was not at least a contributing factor in the rise of internal conflict at Rome.

- "The landing of Spanish troops near Tampico [about 1829] rallied the [Mexican] nation to a unified effort, and the intrepid General Santa Anna... defeated the invaders.... For a moment, the victory bolstered Mexican national pride. But now the danger from abroad that had served to unite the country... vanished and internal dissensions took on a new and ugly face."[34]

- With the disappearance of the external danger from Britain at the end of the American War of Independence in 1783, "disunity began to threaten to turn into disintegration. ... The states were setting up their own tariff barriers against each other and quarreling among themselves... ." This no doubt is why John Adams (the future President) wrote not long after the end of the war that the United States needed an external enemy to protect it from the "danger of dividing."[35]

- During the latter part of World War II, when it had become clear that Germany was irrevocably on the road to defeat, "the Anglo-American accord, which had held very strongly during the testing two and a half years of defeat followed by only peripheral attack, instead of being warmed by the sun of victory began badly to cool. ... [The dispute about Operation Anvil] escalated between 21 June and 1 July [1944] from disagreements at the Chiefs of Staff level to exchanges between Prime Minister and President that were far more acrimonious than anything which had previously passed between them. ... [T]he Anvil disagreement was the beginning of a new pattern. Before it the American and British Chiefs of Staff had rarely disagreed on a major issue. After it they were rarely on the same side of any issue... ."[36]

For further examples see note 164 to Chapter Three, and Beehner's article.[37]

D. In Part II of Chapter Two we discuss self-prop systems that arise to challenge the dominant global self-prop systems. All the examples we give there consist of (formal or informal) organizations of human beings, but self-prop systems that challenge the global self-prop systems also appear at the biological level. Thus there are invasive species—plants or animals that multiply uncontrollably in new environments[38]—and new infectious diseases (e.g., AIDS and Lyme disease) that arise more rapidly than means for curing or preventing them can be found.[39] In addition, older varieties of disease-causing bacteria that once seemed well under control have evolved new forms that are resistant to antibiotics, so that the corresponding diseases are difficult or impossible to cure.[40]

But in the long run these self-prop systems will probably be less dangerous to the global self-prop systems than will those biological self-prop systems that have been intentionally or unintentionally created or altered through direct human action, e.g., through genetic engineering. One would have to be extraordinarily naïve to imagine that organisms created, altered, or manipulated by humans will always remain safely under control, and in fact there already have been cases in which such organisms have *not* remained under control, including cases in which organisms have escaped from research facilities.[41] For example, the so-called "killer bees" are a hybrid of European and African bees that escaped from a research facility in Brazil. Since then they have spread over much of South America and into the United States, and they have killed hundreds of people.[42] Something much, much worse could happen at any time, for the safety record of our biological laboratories is appallingly bad.[43]

It's true that, to date, no biological self-prop system affected by conscious human intervention has come close to threatening the survival of any of the dominant global self-prop systems, but present-day biotechnology is still in its infancy in comparison with what we can expect for the coming decades. As human interventions in biology reach further and further, the risk of disastrous consequences continually rises, and as long as the technological equipment needed for such interventions exists, there are no practicable means of controlling this risk. Small groups of amateurs are already dabbling in genetic engineering.[44] These amateurs wouldn't have to create synthetic life or do anything highly sophisticated in order to bring on a disaster; merely changing a few genes in an existing organism could have catastrophic consequences. The chances of disaster in any one instance may be remote, but there are potentially thousands or millions of amateurs who could begin monkeying with the genes of microorganisms, and thousands or millions of minute risks can add up to a very substantial risk. And the risk has now been vastly increased by the discovery of a powerful new technique that makes gene-editing cheap, quick, and easy.[45]

Some people think it may become possible in the future to create microscopic ("nanotechnological"), non-biological self-prop systems that could reproduce themselves uncontrollably, with deadly consequences for the whole world.[46] Others claim that (macroscopic) self-reproducing robots will probably be built, and even the rabid technophile Ray Kurzweil admits that such machines will evolve beyond the control of human beings.[47] This writer does not have the technical expertise to judge whether such speculations are plausible or whether they should be dismissed as science fiction. Yet, today's science fiction often turns out to be tomorrow's fact.

Because of their ability to reproduce themselves by the billions in a short time, microscopic self-prop systems, biological or not, may prove to be especially dangerous to the global self-prop systems. On the other hand, human self-prop systems may turn out to be more dangerous after all, not only because they are intelligent, but also because they exist as subsystems of the global self-prop systems and therefore can potentially impair the integrity of the latter. But this line of inquiry is leading us too far into speculation, so we'll drop it here.

E. In Part II of Chapter Two we've argued that when only relatively few individuals are available from among which to select the "fittest" (in the Darwinian sense), the process of natural selection will be inefficient in producing self-propagating systems that are fit for survival. We illustrate with an example.

The inefficiency of government agencies or enterprises, in comparison with private enterprises, is notorious, and the reason is clear: Natural selection is not operative among the agencies or enterprises of a given government. If a government-owned or government-controlled agency or enterprise is inefficient—even grossly inefficient—the government tries to reform it in some way, or simply gives it enough money to keep it from collapsing. Rarely indeed will a government allow such an agency or enterprise to die a natural death. In contrast, private enterprises that become inefficient are (barring government interference) eliminated by natural selection.[48]

It seems safe to say that among private enterprises—just as among biological organisms—natural selection leads to the evolution of sophisticated mechanisms that promote the vigor of such enterprises—*including mechanisms that are too complex or subtle to be understood, controlled, or even recognized by human beings.* Students of business administration do of course understand many of the mechanisms at work in successful enterprises. Clearly, however, they are far from a complete understanding of all such mechanisms, for if the principles underlying the efficient functioning of private enterprises were fully understood, then government agencies or enterprises could be made equally efficient by applying to them the same principles. Government agencies and enterprises do try to apply the known principles of business administration, but they nevertheless remain far less efficient than private enterprises—because a great deal of what makes an enterprise efficient remains unknown to, or beyond the control of, human beings.[49]

However, even if natural selection is inoperative among the agencies or enterprises belonging to a given government, natural selection does operate on governments and on the nations they govern. For example, when the countries of the communist bloc failed to compete successfully with the West, their governments and their economic systems were radically transformed in imitation of Western governments and economic systems. The Soviet Union broke apart, and from its fragments new nations under new governments were born. So why doesn't natural selection make national governments, including governmental agencies and enterprises, equal to private enterprises in vigor and efficiency?

In any capitalist system there are many thousands of business enterprises. New enterprises are continually being formed, while some older enterprises go bankrupt, or are absorbed by more powerful enterprises, or are split into two or more separate enterprises. Thus, ample scope for evolution through natural selection is provided by the number of business enterprises and the fluidity with which such enterprises are formed or eliminated. But there are only about two hundred sovereign nations in the world. The creation of new nations and the demise of old ones are infrequent events. Likewise infrequent is the replacement of a nation's government by a new government of a different type. Thus, among nations and their governments, there is only relatively limited scope for evolution through natural selection, and this, we think, explains why governments, with their agencies and enterprises, have not evolved to the same level of efficiency as private enterprises have.

F. One of the most serious mistakes that people make in thinking about the development of societies is to assume that human beings make collective decisions of their own free will and can impose those decisions on their society, as if human volition were something existing *outside* of the organizational structures of society and capable of acting independently of those structures. In reality, human volition is to a very significant extent a *product* of the organizational structures of society,[50] for one of the most important factors that determine the success of an organization is its capacity for *people-management*; that is, its ability to induce people to think and act in ways that serve the needs of the organization.

Some techniques of people-management may be described as "external," meaning that they are used to influence the thought and behavior of people who are not members of the organization that applies the techniques. External techniques include, among others, those of propaganda[51] and public relations. Propaganda and public relations techniques can also be applied internally, to manage the behavior of the members of the organization that applies the techniques; and other techniques are designed specifically for internal use. Business schools give courses in a subject called "Organizational Behavior," which is, in part, the study of techniques through which an organization can manage the behavior of its own members.[52] Also important are techniques for selecting individuals who are suited to become members of a given organization.[53]

But we maintain that the people-managing capability of organizations is not limited to *techniques*, that is, to methods understood and consciously applied by human beings. We argue that through natural selection organizations evolve mechanisms *not* recognized or understood by human beings that tend to induce people to act in ways that serve the needs of the organization. This ties in with what we argued in Part E of this Appendix, about the operation of natural selection among business enterprises.

Of course, all these conscious and unconscious mechanisms put together are very far from achieving complete control over human behavior. The mechanisms are effective only in a statistical sense: They tend *on average* to make people think and act in ways that serve the organizations that possess the mechanisms, but different individuals are influenced in different degrees, and there are always exceptional individuals whose thought and behavior are radically at odds with those that would serve the needs of the organizations in question.

Nevertheless, organizations' capabilities for people-management, whether they are consciously applied techniques or subtly evolved mechanisms unrecognized by humans, are highly important, and people who make naïve statements like, "We [meaning society at large] can choose to stop damaging our environment"—as if the human race had some sort of collective free will—are out of touch with practical reality.[54]

A moment ago we said that, through natural selection, organizations evolve mechanisms not recognized or understood by human beings that tend to induce people to act in ways that serve the needs of the organization.

Let's illustrate with an example.

Until recent times, when technological and economic strength became paramount in warfare, the fighting quality of a society's soldiers was an important factor in the process of natural selection among societies. All else being equal, those societies that produced the best warriors tended to expand their power at the expense of other societies. It's unlikely that military experts would attribute differences in fighting quality solely to causes that are known and controlled by human beings, such as training techniques or methods of military organization. Rather, there are cultural differences among societies—differences that can be identified, if at all, only on a highly speculative basis—that affect the fighting quality of soldiers. Presumably societies have evolved, through natural selection, cultural mechanisms that have tended to produce better soldiers.

Warriors of primitive societies, or of societies at a relatively early stage of civilization, have seldom been able to stand up in pitched battles against trained and experienced European troops, unless the latter were grossly outnumbered, taken by surprise, confused by unfamiliar terrain, or otherwise placed at a grave disadvantage.[55] This cannot be attributed solely to the superiority of European weapons,[56] which indeed have not always been superior under the relevant conditions of combat. Nor can it be attributed to physical courage; if anything, primitives are probably braver on an individual basis than Europeans are.[57] The superiority of European troops can best be attributed to (unidentified) cultural mechanisms evolved through natural selection in the course of millennia, during which European history has been characterized by constant warfare. Of course, there has always been warfare among primitives, too, but such warfare has typically been carried on primarily through guerrilla-like raids rather than pitched battles. So it's not surprising that primitives tend to make excellent guerrilla fighters but are rarely able to put together a regular army capable of facing Europeans on equal terms. Societies at an early stage of civilization, like those of the Aztecs and Incas, ordinarily have had extensive experience of pitched battles, but perhaps have not been subjected to selection through that type of warfare for the same length of time or at the same level of intensity as European societies have; and this may be the reason why their armies have been unable to stand up against European ones.

The fighting qualities of soldiers could be argued ad infinitum, but our interest here is not in fighting qualities per se (nor do we mean to make any value judgment about such qualities). Our purpose at the moment is only to illustrate the point that human organizations evolve, through natural selection, mechanisms that favor their survival and expansion, *including* mechanisms that are not understood or recognized by human beings.

G. In commenting on an earlier, less complete exposition of the theory developed in Chapter Two of this book, Dr. Skrbina observed that a small, isolated island might be considered analogous, for the purposes of the theory, to the Earth as a whole, and he raised by implication the question of whether a counterexample to the theory might be found on a small island without human inhabitants.[58] A proper discussion of this question would require a good knowledge of the biology of small, isolated islands, which this writer does not have. Let's merely take note of the fact that the smaller the island, the less biodiversity it has.[59] This perhaps makes it doubtful whether the ecosystem of such an island could be "highly complex" (as students of industrial accidents use that term); or whether it could be "rich" enough so that (under Proposition 1 of Chapter Two) new self-propagating systems would continually arise to challenge the dominant ones.

So much for islands without human inhabitants. It may be worthwhile, however, to glance briefly at small, isolated islands occupied by humans at a primitive technological level, of which Jared Diamond provides us with two relevant examples: Easter Island and Tikopia. Easter Island certainly offers no counterexample to our theory, since its inhabitants did indeed devastate it as far as was possible with the limited technology at their disposal.[60] Tikopia, on the other hand, merits a closer look.

Tikopia is so tiny (1.8 square miles[61]) that a good runner could doubtless go from one end of the island to the other in somewhere between ten minutes and an hour, depending on the shape of the island, the nature of the terrain, and the straightness or crookedness of the footpaths. Thus, sufficiently rapid transportation and communication were possible between any two parts of Tikopia, and self-prop systems spanning the entire island—analogous to the global self-prop systems considered in Chapter Two—could have developed.

It's impossible to know whether such self-prop systems did in fact develop on Tikopia in the remote past. What we do know is that in the course of their first 800 years on the island the original settlers *did* devastate Tikopia ecologically,[62] but—probably because they had no advanced technology—they apparently didn't devastate it so thoroughly as to cause a major die-off of the human population. Instead, they were able to support themselves by adopting new methods of food production.[63] It's not clear that their economy could be called stable, since they changed it repeatedly over the next 2,000 years until significant European intervention occurred around 1900 AD. But they didn't suffer economic collapse.[64]

The Tikopians moreover seem to have achieved something analogous to the "world peace" considered in Part II of Chapter Two—though it was not entirely stable, as we'll point out in a moment. To the extent that it *was* stable, its stability can be attributed to the fact that Tikopian society was neither highly complex nor tightly coupled, and was not "rich" enough (in the sense of Proposition 1 of Chapter Two) so that new self-prop systems would frequently arise to challenge the island's dominant self-prop systems. The total population of the island was only about 1,300,[65] and within a culturally uniform population of that size we wouldn't necessarily expect any new, strong, aggressive self-propagating human groups to arise within any reasonable period of time.

Even so, the Tikopian "world peace" was not so stable as to prevent all destructive competition: On at least two occasions there were wars in which entire clans were exterminated.[66] Because the Tikopians fought only with primitive weapons (bows and arrows, etc.), their wars damaged only the Tikopians themselves and not their environment. We can imagine what would have happened if they had had advanced technology to fight their wars with; most of us have seen photographs of World War I battlefields ravaged by high-explosive shells, whole forests torn to shreds and so forth.[67] Of course, it's highly unlikely that an island the size of Tikopia could have the mineral resources to sustain an advanced technology. But if it did, then even nonviolent economic competition—even just mining activities alone—would have been enough to ruin the island.

Thus the example of Tikopia does not undercut the theory developed in Chapter Two. Because the islanders lacked advanced technology, and because their society was neither highly complex nor tightly coupled and was not "rich" enough to ensure the frequent emergence (under Proposition 1 of Chapter Two) of vigorous new self-prop systems, Tikopia did not satisfy the conditions for the theory to be applicable.

NOTES

1. LeBlanc, pp. 73–75.
2. Ibid., p. 75.
3. Ibid., p. 73.
4. NEB (2003), Vol. 21, "International Relations," p. 829.
5. Ibid. But it's not clear whether Lenin ever actually made that statement. See Horowitz, p. 152. On the subject of capitalists' trade with the Soviets, NEB, loc. cit., does not seem entirely consistent with Ulam, pp. 196, 265, who says that the Soviets received only "a mere trickle" of help from the West. But Ulam, p. 337, acknowledges that the capitalists were willing at least to some extent to do business with the Soviets, so, lacking a definitive resolution of the apparent inconsistency, we will let this passage stand as in the first edition of the present work.
6. Alinsky, p. 150.
7. The story is told by Peterson and, less completely, by Utt.
8. Peterson, p. 150n6. See also pp. 160–63.
9. Ibid., pp. 151, 167.
10. Ibid., pp. 150–51. Utt, p. 12.
11. NEB (2003), Vol. 21, "International Trade," pp. 900–03.
12. Ibid., p. 905 ("There is general agreement that no modern nation... could really practice self-sufficiency...").
13. Ibid. See also "Relying on China is a big mistake," *The Week*, Oct. 22, 2010, p. 18.
14. See "How supply chains hinge on Asia," *The Week*, Nov. 11, 2011, p. 42.
15. NEB (2003), Vol. 20, "Germany," p. 115; Vol. 21, "International Relations," p. 814; Vol. 29, "War, Theory and Conduct of," p. 652, and "World Wars," pp. 963, 969, 976, 986.
16. Ibid., Vol. 29, "World Wars," pp. 963, 969–970, 976, 977, 979–980, 997–98, 1008. Patterson, p. 121. Dunnigan & Nofi, p. 245.
17. *GMO Quarterly Letter*, April 2011, p. 18. Since GMO is a large investment firm, it is hardly likely to have leftist or radical-environmentalist leanings.
18. Boorstin, pp. 105, 120, 163, 193, 260, 261, 263–65. W.S. Randall, pp. 189, 229.
19. Zimmermann, pp. 266–67. This doesn't necessarily mean that European mining methods were more environmentally sound than American ones.
20. Presumably China is not considered a "developed country." Cf. note 17. China's environmental irresponsibility is so well known that it doesn't seem necessary to cite any authority, but as examples we mention "The cracks in China's engine," *The Week*, Oct. 8, 2010, p. 15; Bradsher, p. A8; *USA Today*, Feb. 25, 2014, p. 2A; March 5, 2015, p. 5A; Dec. 2, 2015, p. 5A; Dec. 8, 2015, p. 3A.
21. Pirenne, e.g., pp. 166–173, 194–95, 236. Elias, pp. 224, 229.

22. *World Book Encyclopedia* (2015), Vol. 15, "Persia, Ancient," p. 297.

23. Ebrey, pp. 64, 70, 85, 116, 141–42, 143 (photo caption), 207, 209, 214. Mote, pp. 17–18, 359, 620–21, 646–653, 714, 749, 903, 917, 946. NEB (2003), Vol. 16, "China," p. 106.

24. NEB (2003), Vol. 29, "War, Technology of," p. 622. Mote, p. 436.

25. Malpass, pp. 68–69. East, p. 160.

26. See Diamond, pp. 164–66.

27. Hassig, p. 51.

28. Ibid., pp. 53, 67.

29. Davies, pp. 46, 110–14, 128, 199–201, 218, 219. Hassig, pp. 11–22, 26, 64, 157, 171–72, 253–54, 256–57.

30. Davies, pp. 183–84, 191, 199–201, 207. Hassig, p. 254.

31. Davies, pp. 107, 110, 112, 128, 201, 204–05, 207, 221. Hassig, pp. 22, 25–26, 195, 198, 229, 231, 263. It should be remembered, however, that Aztec history prior to the arrival of the Spaniards is based on sources of very doubtful reliability. See Davies, p. xiv.

32. Beehner, p. 9A.

33. NEB (2003), Vol. 20, "Greek and Roman Civilizations," p. 300.

34. Bazant, p. 43.

35. NEB (2003), Vol. 29, "United States of America," pp. 216–17. See also McCullough, pp. 397–98. Adams wrote his comment about the need for an external enemy in the margin of a book he was reading. Haraszti, p. 149. From ibid., pp. 140–42, it can be inferred that Adams probably wrote the comment in 1784 and certainly wrote it before Franklin's death in 1790.

36. Jenkins, pp. 748–750.

37. Beehner, loc. cit.

38. E.g., Sodhi, Brook & Bradshaw, p. 516; Weise, "Invasive Species," p. 4A. Examples: Pythons in Florida. *The Week*, Feb. 17, 2012, p. 23. Feral pigs in southwestern U.S. *The Atlantic*, Nov. 2009, p. 22. Kudzu vine in eastern U.S., quagga mussels in Lake Michigan. Invasive species are "a nasty side effect of modern transportation technology," by means of which exotic species are intentionally or unintentionally brought into new environments. "Nature's marauders," *The Week*, Dec. 10, 2010, p. 15. Attempts to control invasive species by introducing nonnative predators tend to backfire because the predators themselves are likely to get out of control. Hamilton, p. 58.

39. "Since the mid-1970s, more than 30 new diseases have emerged… . Most of these are believed to have moved from wildlife to human populations. … Damaged ecosystems—characterized by toxins, degradation of habitat, removal of species and climate change—create conditions for pathogens to move in ways they wouldn't normally move." "Tracking Disease," *Newsweek*, Nov. 14, 2005, p. 46. Once a disease has crossed over to humans from some other species, modern

transportation technology, population density, and urbanization make it possible for the disease to spread widely. Quammen, p. 102. "AIDS in the 19th Century?," *The Week*, Oct. 17, 2008, p. 24. New diseases often are mutated forms of earlier ones. E.g., ibid.; "Mutant rabies is spreading," ibid., May 22, 2009, p. 19. See also *USA Today*, Dec. 18–20, 2015, p. 4A; Jan. 28, 2016, p. 1A; Jan. 29, 2016, p. 3A.

40. E.g.: Allan, p. 34. *The Economist*, April 2, 2011, pp. 73–75. *USA Today*, Oct. 28, 2013, p. 10A; Dec. 17, 2013, pp. 1A–2A; March 5, 2014, p. 6B; Aug. 5, 2015, p. 3A; May 27, 2016, p. 3A.

41. E.g., "Experimental Cotton Seed in Accidental Mix," *Denver Post*, Dec. 4, 2008, p. 13A. See also ibid., Aug. 23, 2005, p. 2B ("Genetically modified wheat pollen can drift to other plants more easily than scientists believed, passing genes to... weeds... .").

42. Blau, especially pp. 16–18. NEB (2003), Vol. 2, "bee," p. 42. *USA Today*, Oct. 9, 2014, p. 5A; Oct. 10, 2014, p. 4A.

43. E.g.: *Denver Post*, Aug. 8, 2007, p. 14A. *USA Today*, March 2, 2015, pp. 1A–2A; May 29–31, 2015, pp. 1A, 4A, 5A; June 4, 2015, p. 1A; June 29, 2015, pp. 1A–2A; July 7, 2015, p. 3A; July 22, 2015, p. 3A; June 3–5, 2016, pp. 1A–2A; Jan. 5, 2017, pp. 1A–2A. Diamond, p. 54.

44. Weise, "DIY Biopunks," p. 7A.

45. Feibus, p. 5B.

46. Joy, pp. 246–48. Keiper, pp. 27–28. See also "A molecular motor," *The Week*, Sept. 23, 2011, p. 23 (reporting nano-sized "motor").

47. Robots of the future "should be able to self-replicate." "What are the odds?," *The Week*, July 2–9, 2010, p. 45 (summarizing an article from *Scientific American*, June 2010).

48. Compare Bowditch, Buono & Stewart, pp. 264–65; Steele, pp. 87–88.

49. From Bowditch, Buono & Stewart, passim, e.g., pp. 31–32, it is clear how far the experts are from a full understanding of what makes an enterprise efficient.

50. See Appendix Two in Kaczynski, Fitch & Madison edition.

51. For information on modern propaganda techniques, see Lindstrom, and also Wu, in our List of Works Cited.

52. See, e.g., Bowditch, Buono & Stewart, passim.

53. Peck, pp. 74–84.

54. For example, Jared Diamond's book is titled *Collapse: How Societies Choose to Fail or Succeed*, as if societies could consciously make choices of that kind.

55. E.g., Davies, pp. 249–250, 271 (military superiority of Spaniards over Aztecs). Ibid., p. 252 ("It was only... by bombarding [the Spaniards] from the rooftops in Tenochtitlan, or from above the deep ravines in Peru, that the Indians were able to achieve a measure of success."). On this subject Hassig does not seem entirely consistent. On pp. 266–67 he says that "the Aztecs were [militarily] a

match for the Spaniards," but also that the Aztec system was "a viable one... in the absence of a major competing power around which disaffected members could unite. But this vacancy was filled by the Spaniards." Given the colossal size of Aztec armies—e.g., 400,000 men (p. 227); 100,000 men (p. 229, p. 233)—a few hundred Spaniards could not have constituted "a major competing power" unless they were militarily more than a match for far larger numbers of Aztecs.

56. E.g., the North American Indians "could not stand up against a bayonet charge," Wissler, p. 93, even though bayonets would have been no more effective than the spears of primitives. Davies, pp. 250–51, discusses the reasons for the Spaniards' military superiority over the Aztecs, including their purportedly superior weapons, and then concludes on p. 252: "The psychological superiority of the Spaniards in the battle-field was probably more decisive than any other factor.... Face to face, the Indians were simply not a match for the Spaniards...." Hassig, pp. 237, 238, agrees that the Spaniards' advantage in weaponry was not the decisive factor in their victory over the Aztecs.

57. E.g., Davies, p. 250 (Spanish chroniclers insisted on the bravery of the Aztecs); p. 277 (referring to "many feats of individual bravery" by Aztecs against Spaniards). Hassig, p. 237, says that in "skill and valor" the "individual Aztec warriors were... the equal of any Spanish soldier... ." Turnbull, *Change and Adaptation*, pp. 89–90, 92, describes traditional Africans' contempt for the cowardice of Europeans.

58. Letter from David Skrbina to the author, Aug. 10, 2011.

59. Edward O. Wilson has "offered a formula that mathematically predicts a geometric reduction in the biodiversity of a given habitat as the size of the habitat shrinks." French, p. 72.

60. Diamond, pp. 79–119.

61. Ibid., p. 286.

62. Ibid., p. 292.

63. Ibid.

64. Ibid.

65. Ibid., p. 289.

66. Ibid., p. 291.

67. "Huge tracts of woodlands were reduced to muddy fields of splintered tree trunks, devoid of wildlife." *Polish American Journal*, March 2015, p. 16.

APPENDIX THREE

Stay on Target

What follows is a heavily rewritten excerpt from a letter to the Editor-in-Chief of the *John Jay Sentinel*, a student newspaper at the John Jay College of Criminal Justice. In its original form the letter was published in the March 2011 and April 2011 issues of the *Sentinel*. The editor had correctly pointed out that economic competition under capitalism encouraged the development of technology, and he asked me whether it would therefore be worthwhile to spend time and effort on eliminating capitalism. Here is my answer:

Those of us who believe that the technological system is an evil are often tempted to attack some of the subordinate evils that are associated with it, such as capitalism, globalization, centralization, bureaucracy, big, intrusive governments, environmental recklessness, and gross economic inequality. This temptation should be resisted. One may, of course, use evils like those I've listed as tools to attack the technological system by pointing out that similar evils inevitably accompany any such system. But it is inadvisable to attack any of the subordinate evils independently of an attack on the technological system as a whole.

What makes the subordinate evils tempting targets for attack is that there already are substantial numbers of people who strongly resent them and could be rallied to resist them; and if any of these evils could be eliminated, the growth of the technological system would be retarded and its negative consequences somewhat mitigated. Capitalism, for instance, is at present the economic system that is most conducive to technological development, so if you could get rid of capitalism you would to some extent slow technological progress; in addition, you would reduce economic inequality. Globalization contributes to economic and technological efficiency because there are obvious advantages to a system in which natural, human, and technical resources can be freely transferred from any one part of the world to any other part where they may be needed. So if you could do away with globalization and isolate each region of the world economically from all the others, technological progress would be significantly slowed. Centralization too is important to technological progress. For example, in order to keep the U.S. economy functioning properly there has to be some central authority to regulate banking, print money, and so forth, otherwise the U.S. would experience the same difficulties as did Germany prior to its unification, when much of the country was still divided into numerous small, independent states, each with its own banking regulations, its own currency, its own weights and measures, etc.[1]

> As many petty states as there were, ... so many were the different civil and criminal codes, so many the different kinds of coins and banknotes, so many the different military, financial, and transportation-related institutions. ... The citizen of Württemberg needed a passport to travel to Baden. For a stay in Koburg-Gotha, Braunschweig, or Schwarzburg-Rudolstadt, the citizen of Baden needed to exchange his money.[2]

For normal economic development, the financial and commercial regulation of Germany had to undergo a centralizing process that spanned most of the 19th century.[3] If centralization could somehow be reversed in Germany—or in the U.S. or any other country—economic growth and technological progress there would be significantly impeded.

So why not attack centralization? First, it would be exceedingly difficult to attack centralization successfully. An organization or a movement would have to concentrate all its energy on that attack, and even if it succeeded in substantially reducing centralization the result would be only to slow technological progress to a certain extent; neither the technological system nor the principal evils associated with it would be eliminated. Thus, in attacking centralization the movement would use its resources inefficiently: It would expend vast energy in the hope of only a modest gain.

Worse still, by concentrating its energy on the campaign against centralization, the movement would distract attention (its own and other people's) from the most important target, which is the technological system itself.

In any case, an attack on centralization could not be successful. Of course, there is no special difficulty about decentralizing in situations where centralization has proven to be economically inefficient. E.g., excessive centralized control over economic activity, otherwise known as socialism, has largely died out due to its inefficiency. But where centralization promotes efficiency, its prevalence is guaranteed by a process of natural selection.[4] Systems that are more centralized (in aspects in which centralization contributes to efficiency) thrive better than those systems that are less centralized; hence, the former tend to expand at the expense of the latter. Since inefficiency imposes economic and other hardships on people, most will oppose decentralization. Even the majority of those who now hold a negative view of centralization would oppose decentralization when they found out what it cost them in terms of efficiency. For example, if you wanted to let each state of the Union establish its own monetary policy and print its own currency independently of all the other states, your proposal would be dismissed as ridiculous. Even if you somehow succeeded in putting such a measure into effect, the negative consequences—monetary chaos and so forth—would outrage so many people that centralized control in monetary matters would soon be reinstated.

Needless to say, if future developments should ever make centralized systems economically and technologically *in*efficient in comparison with less centralized ones, then it will be relatively easy to decentralize. But in that event your attack on centralization will be *promoting* technological progress rather than retarding it. In either case, attacking centralization is not an effective way of resisting technological progress.

Arguments very similar to the foregoing apply to any effort to eliminate capitalism. To have any hope of eliminating capitalism a movement would have to concentrate all its energy on that task, and even if it succeeded in eliminating capitalism the gain would be very modest, because technological progress would continue, though at a somewhat slower rate. There was no capitalism in the Soviet Union, for example, yet that country was by no means a negligible force technologically. Even before World War II the Soviets were among the leaders in nuclear physics;[5] their MiG 15 jet fighter shocked Western forces in the Korean War with its speed and agility;[6] the Soviets were the first to develop a really successful jet airliner, the Tu-104;[7] and the Soviet Union was the first nation to put an artificial satellite into orbit.[8]

Thus, an antitechnological movement that focused on the elimination of capitalism would gain little in return for an enormous expenditure of energy. What is worse, by focusing on capitalism the movement would distract its own and other people's attention from the far more important objective of bringing down the technological system itself.

An attack on capitalism moreover would be futile, or would be successful only temporarily and in a few countries at most. Capitalism has become the world's dominant economic system through a process of natural selection; it has replaced other systems because under present-day conditions capitalism is economically and technologically more efficient. For this reason, even if you could get rid of capitalism in some countries, these would tend strongly to revert to capitalist economic structures as the relative inefficiency of their non-capitalist systems became apparent. This has been demonstrated through experience: When the socialist countries of Eastern Europe couldn't keep up with the West economically or technologically, they adopted capitalist systems. Sweden once was ideologically socialist, but in practical terms socialism never got very far in that country. Today Sweden is still a capitalist welfare-state—and is becoming less of a welfare state as it reduces benefits in the interest of economic efficiency.[9] China remains nominally socialist, but for the sake of economic success the Chinese government now allows a great deal of private enterprise, i.e., capitalism.[10] In Nicaragua the Sandinistas still pretend to be socialist, but in reality they are turning to capitalism.[11] This writer knows of only two countries left in the world that are free of capitalism: Cuba and North Korea. No one wants to imitate Cuba or North Korea, because they are economic failures. And that's why Cuba is now (2011) taking some timid steps in the direction of capitalism.[12]

So it's clear that as long as we live in a technological world we will never get rid of capitalism unless and until it is superseded by some system that is economically and technologically more efficient.

The arguments I've outlined here in reference to centralization and capitalism are equally applicable to globalization, bureaucracy, big, intrusive governments, environmental recklessness, and any number of other evils the elimination of which would merely impair the efficiency of the technological system while still permitting it to grow. As long as society remains saturated with the values of the technological system, most people will not accept any measures that seriously impede the functioning of that system. In order to get people to accept such measures, you would first have to convince them that the supposed "benefits" of modern technology are not worth the price that has to be paid for them. Thus, your ideological attack must be focused on modern technology itself. An attempt to eliminate capitalism, globalization, centralization or any other subordinate evil can only distract attention from the need to eliminate the entire technological system.

NOTES

1. Dorpalen, p. 167. Zimmermann, pp. 8–9. NEB (2003), Vol. 20, "Germany," pp. 106, 111, 113. By "unification" we mean not merely the foundation of the German Empire in 1871, but a process that arguably lasted as long as 93 years, from the changes imposed by the French conquerors in 1807 (ibid., p. 102) to the promulgation of a uniform civil code for the Empire in 1900 (Zimmermann, p. 9).

2. Zimmermann, p.8, quoting one "Löwenthal" without any further indication of the source.

3. See note 1, above, and Tipton (entire article). Tipton argues that historians err when they identify a particular date, e.g., 1834 (creation of the Zollverein—the customs union) or 1871 (foundation of the German Empire), as the point at which German economic development "took off": Quantitative data show that German economic development throughout the period in question was a smoothly continuous process in which no "take-off" points are apparent.

But in places (e.g., pp. 222–23) Tipton seems to argue that centralizing events like the creation of the Zollverein or the foundation of the Empire were unimportant for Germany's economic development. If this is what he means, then his argument has to rest on the assumption that such events could not have been economically important unless they were signaled by an immediate change in the rate of economic growth. And that assumption is clearly unjustified. Among other things, as Tipton himself points out, the changes in economic regulation brought about by the Zollverein and the Empire were developed only over a span of decades: The Zollverein was not fully implemented until 1857 (Tipton, pp. 201, 209), while the economically relevant legislation of the Empire was enacted piecemeal and was not completed until 1897 or even perhaps 1900 (Zimmermann, p. 9; Tipton, p. 209). Moreover, realization of the economic consequences of the changes in regulation required certain developments, such as the construction of railroads (Tipton, pp. 200–01, 205), that could not occur overnight.

Thus, the absence of quantitatively identifiable "take-off" points provides no evidence that the centralization of economic regulation was unimportant for economic growth. Tipton himself notes that "[f]ree movement of resources is important for development" (p. 198), and that "[f]actors of production will be more mobile… in an area without internal tariffs, separate monetary systems, or variations in commercial regulations" (p. 200), from which it logically follows that centralized economic regulation is important for economic development.

4. See Chapter Two of this book.

5. NEB (2003), Vol. 21, "International Relations," p. 858.

6. Ibid., Vol. 8, "MiG," p. 117. See also *Air & Space*, Oct./Nov. 2013, p. 80.

7. Woodall, p. 4. Mellow, pp. 61, 65.

8. NEB (2003), Vol. 19, "Exploration," pp. 47–48.

9. *The Economist*, June 11, 2011, p. 58.

10. The private sector is the most vigorous part of China's economy. *The Economist*, March 12, 2011, pp. 79–80, and June 25, 2011, p. 14 of Special Report ("the dynamism in China's economy is mostly generated by non-state firms"). It's true that massive government intervention has played an important role in building up China's economy, but this has been only a temporary stage that is characteristic of backward countries that are straining to catch up with the fully developed industrial nations. See NEB (2003), Vol. 24, "Modernization and Industrialization," p. 288. In all probability, government intervention in China's economy will become less and less conducive to economic vigor as that country moves beyond the "catch-up" phase.

11. *The Economist*, Aug. 27, 2011, p. 33; Nov. 5, 2011, pp. 47–48.

12. *The Week,* April 29, 2011, p. 8. USA Today, May 10, 2011, p. 6A.

APPENDIX FOUR

The Long-Term Outcome of Geo-Engineering

A. In 2009, a correspondent asked me whether I thought nuclear weapons were the most dangerous aspect of modern technology. What follows is my reply, heavily rewritten.

The most dangerous aspect of modern technology probably is not nuclear weapons. It could plausibly be argued that the remedies for global warming that are likely to be adopted constitute the most dangerous aspect of modern technology.

Nations have a strong incentive to avoid using nuclear weapons, at least on any large scale, because such use would probably be suicidal. This doesn't mean that nuclear war can never happen. On the contrary, the risk of it is very real. But a major nuclear war at least is not a strong probability for the foreseeable future.

On the other hand, it is virtually certain that nations will fail to reduce their emissions of carbon dioxide sufficiently and in time to prevent global warming from becoming disastrous. Instead, global warming will be kept in check through "geo-engineering." This means that the Earth's climate will be artificially managed to keep it within acceptable limits.[1] Of the many tools that have been proposed for management of the Earth's climate, three examples may be mentioned here: (i) Powdered iron can be dumped into the oceans to stimulate the growth of plankton that will absorb carbon dioxide from the atmosphere.[2] (ii) Microbes or other organisms may be genetically engineered to consume atmospheric carbon dioxide.[3] (iii) Carbon dioxide may be pumped into underground reservoirs for permanent storage there.[4]

Any attempt at geo-engineering will entail a grave risk of immediate catastrophe. "Geo-engineering makes the problem of ballistic-missile defense look easy. It has to work the first time, and just right."[5] Novel technological solutions usually have to be corrected repeatedly through trial and error; rarely do they work "the first time, and just right," and that's why people "quite rightly see [geo-engineering] as a scary thing."[6]

But let's assume that geo-engineering does work the first time and just right. Even so, there is every reason to expect that the longer-term consequences will be catastrophic.

First: Attempts to meddle with the environment almost always have unforeseen, undesirable consequences. In order to correct the undesirable consequences, further meddling with the environment is required. This in turn has other unforeseen consequences... and so forth. In trying to solve our problems by tinkering with the environment we just get ourselves deeper and deeper into trouble.

Second: For hundreds of millions of years, natural processes have kept the Earth's climate and the composition of its atmosphere within limits that have allowed the survival and evolution of complex forms of life. Sometimes during this period the climate has varied enough to cause the extinction of numerous species, but it has not become so extreme as to wipe out all of the most complex organisms.

When human beings have taken over the management of the Earth's climate, the natural processes that have kept the climate within livable limits will lose their capacity to perform that function. The climate will then be entirely dependent on human management. Since the Earth's climate is a worldwide phenomenon, it cannot be managed by independent local groups; its management will have to be organized on a worldwide basis and therefore will require rapid, worldwide communication. For this reason among others, management of the Earth's climate will be dependent on technological civilization. Every past civilization has broken down eventually, and modern technological civilization likewise will break down sooner or later. When that happens, the system of human climate-management necessarily will break down too. Because the natural processes that kept the climate within certain limits will be defunct, the Earth's climate can be expected to go haywire. In all probability the Earth will become too hot or too cold for the survival of complex life-forms, or the percentage of oxygen in the atmosphere will sink too low, or the atmosphere will become contaminated with toxic gasses, or some other atmospheric disaster will occur.

Third: When the Earth has a managed climate, maintenance of the technological system will be considered essential for survival because, as has just been pointed out, the breakdown of the technological system will probably lead to radical and fatal disruption of the climate. The elimination of the technological system, through revolution or by any other means, would be almost equivalent to suicide. Because the system will be seen as indispensable for survival, it will be virtually immune to challenge.

The elite of our society—the scientists and engineers, the corporation executives, the government officials and the politicians—are afraid of nuclear war because it would lead to their own destruction. But they will be delighted to see the system that gives them their power and their status become indispensable and therefore immune to any serious challenge. Consequently, while they will make every effort to avoid nuclear war, they will be quite pleased to undertake management of the Earth's climate.

B. Peter Ward, an astrobiologist with NASA, warns us that natural processes will make our planet uninhabitable, perhaps as soon as a mere 500 million years from the present.[7] Most of us would be content to look forward to just one million more years of an inhabitable planet, but Ward apparently is not satisfied even with 500 million, so he advocates a system of geo-engineering that is supposed to keep our climate livable for the next several billion years.[8] Ward acknowledges that the level of global cooperation necessary for the creation of an effective system of geo-engineering may be a "utopian pipe-dream,"[9] but he evidently believes that it is not entirely impossible, since he advocates it. But what is really astonishing is Ward's belief that the system of geo-engineering—and therefore necessarily the worldwide civilization on which it will depend—can survive for *billions of years*.

In Chapter One we remarked on the naiveté of scientists in regard to human affairs. Ward has here provided us with an egregious example of it.[10]

NOTES

1. See, e.g., *Time*, March 24, 2008, p. 50.
2. Wood, p. 73, col. 2.
3. Leslie, p. 6, col. 4 (microbes). Wood, p. 73, col. 1 (trees).
4. Wood, p. 73, col. 2. Sarewitz & Pielke, p. 59, col. 3. It necessarily remains an open question whether the carbon dioxide will remain underground as long as the proponents of this plan believe. Even if a "demonstration project" (ibid.) keeps the CO_2 underground for as long as, say, ten years, that doesn't guarantee that it will stay there for a hundred or a thousand years. Moreover, any demonstration project will be carried out with special care by highly qualified experts. But once the procedure becomes routine and is widely applied, there *inevitably* will be

negligence, incompetence, and dishonesty in its execution. Compare, e.g., the *USA Today* articles cited in note 67 to Chapter Two, above.

5. Wood, p. 76, col. 1, quoting Raymond Pierrehumbert, a geophysicist at the University of Chicago.

6. Ibid. See also *USA Today*, Feb. 16, 2015, p. 7A.

7. Ward, pp. 141–42.

8. Ibid., pp. 143, 149.

9. Ibid., p. 143.

10. For another egregious example, see K. Brower, pp. 60, 62.

APPENDIX FIVE

Thurston's View of Stalin's Terror. State Terrorism in General.

A. Stalin's Terror as portrayed by Thurston provides an important example in support of our argument that the power even of "absolute" dictators is in reality far from absolute, so it is worthwhile to point out that, for our purposes, Thurston's revisionist portrayal of the Terror is not seriously inconsistent with the traditional view of Stalin as the "mastermind of a plot to subdue the party and the nation."[1]

We will take Ulam's biography of Stalin to represent the traditional view, but first we have to note three points: (i) Much of what Ulam says about Stalin's motives and intentions can be discounted as rank speculation. Ulam repeatedly indulges in "mind-reading;" without offering any supporting evidence, he tries to tell us what was going on in Stalin's head.[2] In some passages Ulam's book even reads like a novel.[3] (ii) Thurston's statement that Stalin "did not plan the Terror" has to be understood to mean only that Stalin did not plan the Terror *as it actually developed*; Thurston nowhere demonstrates that Stalin could not have planned to initiate a terror campaign of *some* kind. (iii) When Thurston says that the effects of the Terror were largely confined to the elite,[4] the term "elite" has to be understood to include all those who worked with their heads rather than their hands and whose work required a good deal of education or special training.[5]

Now, when Ulam insists that Stalin was "in the main, firmly in control of the purge,"[6] what can he mean? Well, he writes that Yezhov (head of the NKVD,[7] later known as the KGB) "took no important step without obtaining Stalin's sanction. The lists of important people to be shot or otherwise repressed were sent to Stalin by the Commissar for his approval. During the period 1937–39, 383 such lists were submitted to Stalin… ."[8] If we make the modest assumption that the lists contained on average about ten names, then Stalin during the given period would have had to decide the fates of approximately 3,800 people. Could Stalin of his own personal knowledge have been sufficiently familiar with the histories of 3,800 individuals to decide their respective fates rationally? It seems doubtful. More likely Stalin would have had to rely in the majority of cases on information appended to the lists by the NKVD. Thus it would have been really the NKVD, or Yezhov personally, who decided the fates of the listed individuals by choosing the information to be provided to Stalin. Even if we suppose that the lists contained an average of 500 names and make the wildly improbable assumption that Stalin had enough independent knowledge of each of the listed individuals to decide his or her fate rationally, still the lists accounted for only about 190 thousand individuals. But *millions* were executed;[9] what about all the others? Clearly Stalin was in control of only a fraction—probably only a minute fraction—of all the executions that were carried out. Even among that tiny fraction, Ulam admits that there "were excesses and mistakes… even from [Stalin's] point of view—people whom he would have preserved had he known the full circumstances; in some cases his subordinates were settling personal scores… ."[10] Among the vast majority of the executed or imprisoned individuals—those whose fates were *not* decided by Stalin personally—there must have been a far larger proportion who were the victims of "excesses," "mistakes," or personal grudges. How could it have been otherwise, when people were being executed literally by the millions?

So again we have to ask, what can Ulam mean when he says that Stalin was "firmly in control of the purge"? Does he mean merely that Stalin *intended* indiscriminate and apparently senseless executions to take place on a massive scale? Ulam seems to say exactly that.[11] Yet Ulam himself also suggests a different hypothesis: "in 1936 [Stalin] may have desired to strike out just the leaders of potential treason: those few thousand Party officials who in the past had been connected with his rivals. But the mechanics of terror… soon acquired its own momentum."[12]

Even if Stalin did intend the purge to be indiscriminate and massive in scale, Ulam shows how Stalin was manipulated by his subordinates, who "discovered" new plots and treason in order to "demonstrate their zeal and loyalty."[13] There was a certain degree of antagonism between the army and the NKVD, as a result of which the NKVD manipulated Stalin into extending the purge to the army.[14] "There can be no doubt," says Ulam, that Stalin "came to believe in the *essential* veracity of the fantastic tales of treason and sabotage woven by his servants… ."[15] Ulam further writes that Stalin's "feelings must have been those of what might be called controlled panic," and he refers to "Stalin's thrashing about amidst his terrible and contradictory fears… ."[16] Does Ulam have a firm factual basis for these inferences about Stalin's beliefs, feelings, and fears, or is he merely indulging in "mind-reading"? If he is indulging in mind-reading, then his reading of Stalin's mind agrees very well with Thurston's reading of it. So it is difficult to find any major contradiction[17] between Thurston's view of Stalin's Terror and the traditional view as represented by Ulam. The difference between the two seems to be largely a matter of rhetoric.

Ulam provides further evidence that Stalin did not have the Terror under rational control: In 1938–39 Stalin himself concluded that the Terror had gotten out of hand, and he tried to "reassure the 'little people'... that while severe measures would continue to be applied against bigwigs, indiscriminate terror as far as the masses were concerned was a thing of the past."[18] During World War II no terror would have been necessary, because the struggle against a fearsome external enemy united all Russians behind their Leader. But when Stalin resumed the practice of terror after the war he did so on a much smaller scale,[19] presumably because he "did not want and could not afford a repetition" of the "chaos... of 1937–39."[20] True, Ulam suggests that during the last year of his life Stalin may have been planning to resume terror on a mass basis,[21] but only in desperation, because he feared that in old age he was losing his grip on power.[22]

B. A proper discussion of state terrorism in general would be beyond the scope of this book, but I do want to address briefly Thurston's claim that a "system of terror as described by theorists and other scholars has probably never existed."[23] Thurston does not clearly explain why he thinks that Stalin's "Terror" fails to qualify as a "system of terror," but probably he has in mind his own argument that "extensive fear did not exist in the USSR at any time in the late 1930s. ... The sense that anyone could be next, the underpinning of theories on systems of terror, rarely appears."[24] Thurston must mean that "extensive fear," etc. did not exist *throughout the general population*. He could hardly deny that among the high elite—the class of upper-level officials who for the most part were exterminated by Stalin[25]— there had to be "extensive fear" and a "sense that anyone could be next."

Thurston very likely is right in maintaining that "extensive fear" did not permeate the population as a whole. Fischer's account—based on personal experience—of life under Stalin suggests that the working class was largely immune from terror.[26] But this by no means disqualifies Stalin's system as a system of terror. Such a system does not have to be applied to an entire population; it can be limited to some segment of the population.[27] However, during the late 1930s under Stalin it appears that the terrorized segment included not only the high elite, but also the elite in the broader sense described in Part A of this appendix, item (iii).[28]

Thurston tries to dismiss E.V. Walter's study of the system of terror practiced by the Zulu emperor Shaka (also spelled Chaka) by suggesting that the British witnesses on whose accounts Walter relied "may not have been in a position to understand what they saw."[29] But Thurston's suggestion is not credible. Walter's conclusions are supported by the accounts of several British observers,[30] the best of whom, H.F. Fynn, acquired "a thorough knowledge of Zulu language and customs,"[31] and the violent events described were of such a nature that their basic import could hardly be misunderstood, even if there were nuances that the observers overlooked.

The Argentine caudillo Juan Facundo Quiroga and his (so to speak) successor Juan Manuel de Rosas both made use of clearly-defined systems of terror.[32] Other examples could no doubt be identified—very likely in imperial China, for instance,[33] or among 20th-century Latin American dictators—though Thurston is probably right in stating that the "model does not fit Nazi Germany,"[34] and Sarmiento discounts the Reign of Terror of the French Revolution.[35] I would like to suggest, however, that the reign of Henry VIII in England could very possibly be considered terroristic. Certainly Henry's system was haphazard and of low intensity compared with that of Stalin, Shaka, Facundo Quiroga or Rosas, and the terrorized class—that of the courtiers—comprised only a very small fraction of the general population. But arguably the essential elements of a system of terror were there:

- The king consciously used fear as an instrument of governance: "He ruled on the precept that fear engenders obedience."[36]
- Arrests and executions, often of completely innocent persons, tended to be irrational and unpredictable, and sometimes resulted from the whisperings of informers or slanderers seeking to eliminate their rivals or take revenge on their enemies.[37]
- Consequently there was "extensive fear,"[38] a "sense that anyone could be next."[39]
- Innocent victims about to be executed at the king's order often abased themselves, declaring their loyalty to and love for the tyrant who was murdering them.[40]
- As a "legitimate" sovereign, Henry VIII possessed a ready-made

"cult of personality,"[41] but he calculatedly intensified the cult.[42] As with the cults of Stalin and Rosas, Henry's cult of personality relied in part on ubiquitous images of himself.[43]

• Despite his cruelty and injustice, Henry VIII was vastly admired, perhaps worshipped, even by some (or most?) members of the terrorized class.[44]

NOTES

1. See Thurston, p. 17.

2. Ulam, e.g., last paragraph on p. 311 through first five lines on p. 312; first complete paragraph on p. 529; pp. 534–35.

3. Ibid., e.g., last six lines on p. 272 through first two lines on p. 274; last six lines on p. 534 through p. 535.

4. Thurston, pp. 144–150.

5. Ibid., pp. 148 ("People with more education were certainly more likely than others to be arrested."); 149 (evidence that any engineer ran a high risk of being arrested). Fischer, e.g., pp. 149–151, 163, 201, 205, 222, 228–29. Fischer, p. 150, adds that "even factory workers" might be arrested. This is consistent with Thurston, p. 193: "Occasionally [workers] went… into the gulag or to their deaths."

6. Ulam, p. 445.

7. See ibid., pp. 419–420.

8. Ibid., p. 444.

9. Millions were executed even in Thurston's view; "traditional" estimates of the number executed were much higher. See Thurston, pp. xvii, 139–140.

10. Ulam, p. 444.

11. Ibid., pp. 399, 438.

12. Ibid., p. 408.

13. Ibid., pp. 395–98. See also p. 488 (Stalin manipulated by Beria).

14. Ibid., pp. 451–52.

15. Ibid., p. 412.

16. Ibid., p. 457. See also p. 477 (referring to Stalin's "panic that had unleashed terror…").

17. Meaning a major contradiction *for our purposes*, bearing in mind that our purpose here is not to estimate the number of victims of the Terror or anything of that sort, but only to determine whether Stalin was able to keep the Terror under rational control.

18. Ulam, pp. 474–76, 487–88; specifically p. 476.

19. Ibid., pp. 643, 674.

20. Ibid., p. 727.

21. Ibid., pp. 737–38.

22. See ibid., pp. 724–739.

23. Thurston, p. 232.

24. Ibid., p. 159. It's hard to reconcile this with Thurston's own statement that an "atmosphere of panic had set in...," ibid., p. 90.

25. For the extermination of most of the high elite, see Ulam, e.g., pp. 430–31, 438, 441, 447–48, 489. Thurston does not deny the occurrence of this bloodbath among the high elite. See Thurston, p. 68.

26. See Fischer, pp. 151–52, 163–65, 208–09.

27. See Walter, pp. 6–7.

28. See Note 5, above.

29. Thurston, pp. 232–33.

30. Walter, p. 128.

31. Ibid., p. 130.

32. See the works of Sarmiento and of John Lynch that appear in our List of Works Cited; also Ternavasio, pp. 66–73; González Bernaldo, pp. 199–204.

33. Mote, pp. 572–582. Ebrey, pp. 192–93.

34. Thurston, p. 232. The Nazis no doubt used terroristic methods in the concentration camps and in occupied countries, but terror does not seem to have been used in governing Germany itself, at least not before July 20, 1944. It's not even clear that the Nazi regime qualifies as fully totalitarian, given that the German press preserved some degree of independence, Rothfels, p. 49, Skidelsky, p. 254, and the Nazis sometimes tolerated behavior that under Stalin would have meant swift and certain death. E.g., the authorities did not intervene when the Bishop of Münster publicly preached against the "criminal methods" of the Nazi regime, Rothfels, pp. 58–59. Needless to say, the foregoing remarks are not intended in any way to minimize the extraordinary viciousness of the Nazis.

35. Sarmiento, p. 261.

36. Weir, p. 430.

37. E.g., Fraser, pp. 295, 323–24, 336, 342, 392; Weir, pp. 356–57, 368–69, 373, 426–27, 430, 441, 488–89.

38. E.g., Fraser, pp. 272, 389, 393; Weir, pp. 430, 482, 484.

39. Weir, p. 371.

40. E.g., Fraser, pp. 249–253, 255, 257, 353.

41. Weir, pp. 21–22.

42. Ibid., p. 348.

43. Ibid., pp. 349–350, 410, 473.

44. Ibid., pp. 427, 494–95. NEB (2003), Vol. 29, "United Kingdom," p. 51 ("the French ambassador announced that [Henry VIII] was... an idol to be worshiped...").

APPENDIX SIX

The Teachings of Jesus Christ and Their Effect on Society

A. Because of the need for brevity, our treatment of Jesus's teachings in the discussion of Postulate 2 in Part II of Chapter Three left out of account the fact that no one knows for certain what Jesus actually taught.[1] For our purposes, however, this is not important, because our concern is not with Jesus himself but with the question of whether any body of teachings, unsupported by practical action (such as the building of powerful organizations committed to those teachings), can be effective in guiding human behavior on a mass basis. Hence, what matters for our purposes is what the early Christians *believed* the teachings of Jesus to have been. Were the teachings they believed to have been those of Jesus effective in guiding human behavior on a mass basis?

There is a further complication: The early Christians didn't all believe exactly the same things about Jesus. There originally were something like twenty gospels,[2] and even as to the four canonical gospels (Matthew, Mark, Luke, and John) the "evidence suggests that in the early church each congregation would have had its own gospel—exposure to all four might not have been typical until at least the end of the second century."[3] This however does not pose a serious problem for us, because the four canonical gospels presumably represent, to a reasonable approximation, what the main currents in early Christianity believed the teachings of Jesus to have been, and as we saw in Part II of Chapter Three, the canonical gospels proved ineffective in guiding human behavior on a mass basis.

This was true even though Christians did eventually form powerful organizations, mainly the Catholic and Eastern Orthodox churches. These were committed to the propagation of certain religious doctrines—and, of course, to building their own power. In this they were successful. They were also successful to a limited extent in molding human behavior. But what matters for us here is the fact that human behavior was not molded in such a way as to make any substantial fraction of the population of the world's Christian lands behave in accord with the original teachings of Jesus (or what the majority of early Christians believed to be his teachings).

B. To the evidence offered in Part II of Chapter Three, the following can be added:

- The Gospels condemn adultery and fornication,[4] yet an edict of (Pope) Callixtus I (approximately 217–222 AD) gave the Church's approval to sexual relationships between upper-class Roman matrons and their male slaves.[5] It's true that Callixtus I was a maverick, and his action was by no means typical for Christian clergy of the time, but he would hardly have issued his edict if there hadn't been significant numbers of Christian matrons who were already involved in such relationships and therefore were committing fornication if not adultery. No one having even a superficial acquaintance with the history of Europe in later centuries will claim that the teachings of the Gospels have had any substantial effect on adultery and fornication,[6] unless among a tiny minority of rigorists and ascetics.
- At some time between 195 and 240 AD, the distinguished Christian writer Tertullian railed against rich and noble Christian women who went around all dolled up and defended their adornments on the ground that they would have been conspicuous as Christians if they had dressed otherwise.[7] As to female adornment in later times, it is hardly necessary to comment.

C. For whatever it's worth, we note that what we've argued here does not constitute an attack on Christianity as a religion. Christian doctrine—if this writer understands it correctly—holds that Jesus came not to change the way affairs were conducted in this world, but to provide human beings with a path to salvation in some future life.

NOTES

1. Freeman, pp. 19–30.
2. Ibid., pp. 20–21. See also pp. 97–99.
3. Ibid., p. 73.
4. E.g., Matthew 5:27 & 32, 15:19; Mark 7:21; Luke 18:20. See also I Corinthians 7:2, and Augustine, II.3.7, p. 28; II.6.14, pp. 33–34; X.30.41, p. 223.
5. Harnack, p. 210. Callixtus I was Bishop of Rome but not technically a pope, because the Bishops of Rome at that time were not yet titled "Pope." See Freeman, pp. 315–16.
6. See, e.g., Elias, pp. 154–55.
7. Harnack, p. 62n3. See Freeman, p. 180; Isaiah 3:16-24.

LIST OF WORKS CITED

Acohido, Byron, "Hactivist group seeks 'satisfaction,'" *USA Today*, June 20, 2011.

Acohido, Byron, "LulzSec's gone, but its effect lives on," *USA Today*, June 28, 2011.

Acohido, Byron, "Hackers mine ad strategies for tools," *USA Today*, July 16, 2013.

Acohido, Byron, and Peter Eisler, "How a low-level insider could steal from NSA," *USA Today*, June 12, 2013.

Aditya Batra. See Batra.

Agüero. See Conte Agüero.

Agustín, José, *Tragicomedia Mexicana*, Colección Espejo de México, Editorial Planeta Mexicana, Mexico City; Vol. 1, fifth printing, 1992; Vol. 2, 1993.

Alinsky, Saul D., *Rules for Radicals: A Pragmatic Primer for Realistic Radicals*, Vintage Books, Random House, New York, 1989.

Allan, Nicole, "We're Running Out of Antibiotics," *The Atlantic*, March 2014.

Ashford, Nicholas A., and Ralph P. Hall, *Technology, Globalization, and Sustainable Development: Transforming the Industrial State*, Yale University Press, New Haven, Connecticut, 2011.

Astor, Gerald, *The Greatest War: Americans in Combat 1941–1945*, Presidio Press, Novato, California, 1999.

Augustine (Aurelius Augustinus), Saint, *The Confessions*, trans. by Maria Boulding, First Edition, Vintage Spiritual Classics, Random House, New York, 1998.

Azorín (José Martínez Ruiz), *El Político*, Primera edición, Fondo de Cultura Económica, Mexico City, 1998.

Barbour. See Duncan.

Barja, César, *Libros y Autores Clásicos,* Vermont Printing Company, Brattleboro, Vermont, 1922.

Barrow, Geoffrey W.S., *Robert Bruce & The Community of the Realm of Scotland*, Third Edition, Edinburgh University Press, Edinburgh, 1988, reprinted 1999.

Batra, Aditya, "A revolution gone awry," *Down to Earth*, May 16–31, 2011, pp. 23–25. *Down to Earth* is a magazine published in print by the Society for Environmental Communications, New Delhi, India. It is also available at http://www.downtoearth.org.in

Bazant, Jan, *A Concise History of Mexico: From Hidalgo to Cárdenas, 1805–1940*, Cambridge University Press, Cambridge, U.K., 1977.

Beatty, Thomas J., et al., "An obligately photosynthetic bacterial anaerobe from a deep-sea hydrothermal vent," *Proceedings of the National Academy of Sciences U.S.A.*, Vol. 102, No. 26, June 28, 2005, pp. 9306–9310.

Beehner, Lionel, "History warns of Iraq's fall," *USA Today*, July 10, 2014.

Benton, Michael J., "Instant Expert 9: Mass Extinctions," *New Scientist*, Vol. 209, No. 2802, March 5, 2011, pp. i–viii. Benton's article is a pull-out that appears between pages 32 and 33 of this issue of *New Scientist*.

Bernaldo. See González Bernaldo.

Blau, Melinda, *Killer Bees*, Steck-Vaughn Publishers, Austin, Texas, 1992.

Bolívar. See Soriano.

Boorstin, Daniel J., *The Americans: The Colonial Experience*, Phoenix Press, London, 2000.

Botz. See La Botz.

Bourne, Joel K., Jr., "The End of Plenty," *National Geographic*, June 2009.

Bouwsma, William J., *John Calvin: A Sixteenth Century Portrait*, Paperback Edition, Oxford University Press, New York, 1989.

Bowditch, James L., Anthony F. Buono, and Marcus M. Stewart, *A Primer on Organizational Behavior*, Seventh Edition, John Wiley & Sons, Hoboken, New Jersey, 2008.

Bradsher, Keith, " 'Social Risk' Test Ordered By China for Big Projects," *New York Times International*, Nov. 13, 2012.

Brathwait, Richard, *English Gentleman*,1630.

Brooking, Emerson T., and P.W. Singer, "War Goes Viral: How Social Media is Being Weaponized," *The Atlantic*, Nov. 2016.

Brower, David, "Foreword," in Wilkinson.

Brower, Kenneth, "The Danger of Cosmic Genius," *The Atlantic*, Dec. 2010.

Browning, William Ernst (ed.), *The Poems of Jonathan Swift, D.D.*, Vol. I, G. Bell and Sons, London, 1910.

Buchanan, Scott (ed.), *The Portable Plato*, trans. by Benjamin Jowett, Penguin Books, New York, 1977.

Buckley, F.H., "How Machiavelli made Trump into a virtue," *USA Today*, June 1, 2016.

Buechler, Steven M., *Understanding Social Movements: Theories from the Classical Era to the Present*, Routledge, New York, 2016.

Buhle, Paul, and Edmund B. Sullivan, *Images of American Radicalism*, Second Edition, The Christopher Publishing House, Hanover, Massachusetts, 1999.

Bury, J.B., *The Idea of Progress: An Inquiry into its Origin and Growth*, Dover Publications, New York, 1955.

Caputo, Philip, "The Border of Madness," *The Atlantic*, Dec. 2009.

Carr, Nicholas, "The Great Forgetting," *The Atlantic*, Nov. 2013.

Carrillo, Santiago, *Eurocomunismo y Estado*, Editorial Crítica, Grupo Editorial Grijalbo, Barcelona, 1977.

Carroll, Chris, "Small Town Nukes," *National Geographic*, March 2010.

Cebrián, José Luis, et al., *La Segunda Guerra Mundial: 50 años después*, No. 67, "Operación Walkiria. Objetivo: matar al Führer," Prensa Española, Madrid, 1989.

Chernow, Ron, *Alexander Hamilton*, Penguin Books, New York, 2004.

Christian, Brian, "Mind vs. Machine," *The Atlantic,* March 2011.

Christman, Henry M. (ed.), *Essential Works of Lenin*, Bantam Books, New York, 1966.

Churchill, Winston, *A History of the English-Speaking Peoples*, Vol. Four, *The Great Democracies*, Bantam Books, New York, 1963.

Conte Agüero, Luis, *Cartas del Presidio*, Editorial Lex, Havana, 1959.

Coon, Carleton S., *The Hunting Peoples*, Little, Brown and Company, Boston, 1971.

Currey, Cecil B., *Road to Revolution: Benjamin Franklin in England, 1765–1775*, Anchor Books, Doubleday, Garden City, New York, 1968.

Davidson, Adam, "Making it in America," *The Atlantic*, Jan./Feb. 2012.

Davies, Nigel, *The Aztecs: A History*, University of Oklahoma Press, Norman, Oklahoma, 1980, fourth printing, 1989.

De Gaulle. See Gaulle.

De Tocqueville. See Tocqueville.

Dennett, Daniel C., *Darwin's Dangerous Idea*, Simon & Schuster, New York, 1995.

Diamond, Jared, *Collapse: How Societies Choose to Fail or Succeed*, Penguin Books, New York, 2011.

Dimitrov, Georgi, *The United Front*, International Publishers, New York, 1938.

Di Tella. See Tella.

Dorpalen, Andreas, *German History in Marxist Perspective: The East German Approach*, Wayne State University Press, Detroit, 1988.

Drehle, David von, "The Little State That Could," *Time*, Dec. 5, 2011.

Dulles, Foster Rhea, *Labor in America: A History*, Third Edition, AHM Publishing Corporation, Northbrook, Illinois, 1966.

Duncan, A.A.M. (ed.), *John Barbour's The Bruce*, Canongate Books, Edinburgh, 1997.

Dunnigan, James F., and Albert A. Nofi, *The Pacific War Encyclopedia*, Checkmark Books, an imprint of Facts on File, Inc., 1998.

Duxbury, Alyn C. and Alison B., *An Introduction to the World's Oceans*, Third Edition, Wm. C. Brown Publishers, Dubuque, Iowa, 1991.

East, W. Gordon, *The Geography Behind History*, W.W. Norton & Company, New York, 1999.

Ebrey, Patricia Buckley, *The Cambridge Illustrated History of China*, First Paperback Edition, Cambridge University Press, Cambridge, U.K., 1999, ninth printing, 2007.

Ejaz, Sohail, et al., "Endocrine Disrupting Pesticides: A Leading Cause of Cancer Among Rural People in Pakistan," *Experimental Oncology*, June 2004, Vol. 26, No. 2, pp. 98–105.

Elias, Norbert, *The Civilizing Process*, trans. by Edmund Jephcott, Revised Edition, Blackwell Publishing, Malden, Massachusetts, 2000.

Emerson, Ralph Waldo, *Self-Reliance and Other Essays*, Dover Publications, New York, 1993.

Engels, Friedrich, Letter to Joseph Bloch, Sept. 21–22, 1890, in *Der sozialistische Akademiker*, 1. Jahrgang, Nummer 19, Oct. 1, 1895. As we have it, the letter was downloaded from *Das Elektronische Archiv*, http://www.dearchiv.de/php/dok.php?archiv=mew&brett=MEW037&f…

Feeney, John, "Agriculture: Ending the World as We Know It," *The Zephyr*, Aug.–Sept. 2010.

Feibus, Mike, "Are we ready to play God?," *USA Today*, July 24, 2017.

Fischer, Markoosha, *My Lives in Russia,* First Edition, Harper & Brothers Publishers, New York, 1944.

Fluharty, V.L., *Dance of the Millions: Military Rule and the Social Revolution in Colombia (1930–1956)*, University of Pittsburgh Press, 1957.

Foer, Franklin, "Mexico's Revenge," *The Atlantic*, May 2017.

Folger, Tim, "The Secret Ingredients of Everything," *National Geographic*, June 2011.

Foroohar, Rana, "What Happened to Upward Mobility?," *Time*, Nov. 14, 2011.

Foroohar, Rana, "Companies Are the New Countries," *Time*, Feb. 13, 2012.

Fountain, Henry, "A Dream Machine," *New York Times*, "Science Times" section, Late Edition (East Coast), March 28, 2017.

Fraser, Antonia, *The Wives of Henry VIII*, Vintage Books, Random House, New York, 1994.

Freeman, Charles, *A New History of Early Christianity*, Paperback Edition, Yale University Press, New Haven, Connecticut, 2011.

French, Howard W., "E.O. Wilson's Theory of Everything," *The Atlantic*, Nov. 2011.

Fukuyama, Francis, "The End of History?," *The National Interest*, No. 16, Summer 1989, pp. 3–18.

Gallagher, Matt, "No Longer a Soldier," *The Week*, Feb. 11, 2011.

García, Antonio, *Gaitán y el problema de la Revolución Colombiana*, Cooperativa de Artes Gráficas, Bogotá, 1955.

Gardner, Gary, Tom Prugh, and Michael Renner (project directors), *State of the World 2015: Confronting Hidden Threats to Sustainability*, Worldwatch Institute, Island Press, Washington, D.C., 2015.

Gastrow, Peter, *Termites at Work: Transnational Organized Crime and State Erosion in Kenya*, International Peace Institute, New York, Sept. 2011.

Gastrow, Peter, *Termites at Work: A Report on Transnational Organized Crime and State Erosion in Kenya—Comprehensive Research Findings*, International Peace Institute, New York, Dec. 2011.

Gaulle, Charles de, *The Complete War Memoirs of Charles de Gaulle*, trans. by Jonathan Griffin and Richard Howard, Carroll & Graf Publishers, New York, 1998.

Gilbert, Martin, *The European Powers, 1900–1945*, Phoenix Press, London, 2002.

Gilbert, Martin, *The Second World War: A Complete History*, Revised Edition, Henry Holt and Company, New York, 2004.

Glendinning, Chellis, "Notes Toward a Neo-Luddite Manifesto," *Utne Reader*, March/April 1990. Glendinning's article is reprinted in Skrbina, pp. 275–78. The citation to the original source of the article was provided by Dr. Skrbina (letter to this writer, June 28, 2010). I have not seen Glendinning's article as it originally appeared in *Utne Reader*, but only as reprinted in Skrbina.

Goldberg, Jeffrey, "Monarch in the Middle," *The Atlantic*, April 2013.

González Bernaldo, Pilar, "Sociabilidad, espacio urbano y politización en la ciudad de Buenos Aires (1820–1852)," in Sabato & Lettieri, pp. 191–204.

Graham, Hugh Davis, and Ted Robert Gurr (eds.), *Violence in America: Historical and Comparative Perspectives*, Bantam Books, New York, 1970.

Grossman, Lev, "Singularity," *Time*, Feb. 21, 2011.

Guevara, Ernesto "Che," *Diario de Bolivia*, Ediciones B, Barcelona, 1996.

Guillermoprieto, Alma, *Al pie de un volcán te escribo: Crónicas latinoamericanas*, trans. from English by Alma Guillermoprieto and Hernando Valencia Goelkel, Grupo Editorial Norma, Santafé de Bogotá, Colombia, 1995.

Guillette, Elizabeth A., et al., "An Anthropological Approach to the Evaluation of Preschool Children Exposed to Pesticides in Mexico," *Environmental Health Perspectives,* Vol. 106, No. 6, June 1998, pp. 347–353.

Hamilton, Anita, "The Bug That's Eating America," *Time*, July 4, 2011.

Hammer, Joshua, "Getting Past the Troubles," *Smithsonian* magazine, March 2009.

Haraszti, Zoltán, *John Adams & The Prophets of Progress*, The Universal Library, Grosset & Dunlap, New York, 1964.

Harford, Tim, "What Nuclear Reactor Can Teach Us About the Economy," *Financial Times*, Jan. 15, 2011.

Harnack, Adolf (von), *Die Mission und Ausbreitung des Christentums in den ersten drei Jahrhunderten*, zweite neu durchgearbeitete Auflage, II. Band, *Die Verbreitung,* J.C. Hinrichs'sche Buchhandlung, Leipzig, 1906. The copy referenced here is a facsimile put out by the University of Michigan Library, 2016.

Hassig, Ross, *Aztec Warfare: Imperial Expansion and Political Control*, University of Oklahoma Press, Norman, Oklahoma, 1995.

Hayes, Christal, "Immigrants' home life makes trip worth the risk," *USA Today*, June 26, 2018.

Heilbroner, Robert, and Aaron Singer, *The Economic Transformation of America Since 1865*, Harcourt Brace College Publishers, Fort Worth, Texas, 1994.

Hernandez, Rebecca R., et al., "Solar energy development impacts on land cover change and protected areas," *Proceedings of the National Academy of Sciences U.S.A.*, Vol. 112, No. 44, Nov. 3, 2015, pp. 13579–13584. See the correction in the same journal , Vol. 113, No. 12, March 22, 2016, p. E1768.

Hitler, Adolph [sic], *Mein Kampf*, Houghton Mifflin, Boston, 1943.

Hoffer, Eric, *The True Believer*, Harper Perennial, Harper Collins, New York, 1989.

Horowitz, Irving Louis, *El Comunismo Cubano: 1959–1979*, trans. from English by Noevia Lugones and Rubén Miranda, Biblioteca Cubana Contemporánea, Editorial Playor, Madrid, 1978/79.

Hoyle, Fred, *Of Men and Galaxies*, University of Washington Press, Seattle, 1964.

Huenefeld, John, *The Community Activist's Handbook*, Beacon Press, Boston, 1970.

Illich, Ivan, *Tools for Conviviality*, Harper & Row, New York, 1973.

Isaacson, Walter, *Kissinger: A Biography*, Simon & Schuster Paperbacks, New York, 2005.

Ivey, Bill, *Handmaking America: A Back-to-Basics Pathway to a Revitalized American Democracy*, Counterpoint, Berkeley, California, 2012.

Jacobson, Douglas W., "The Founding of Zegota," *Polish American Journal*, Sept. 2011.

Jenkins, Roy, *Churchill: A Biography*, Plume, a member of Penguin Putnam, Inc., New York, 2002.

Johnson, George, "The Nuclear Tourist," *National Geographic*, Oct. 2014.

Johnson, Keith, and Russell Gold, "U.S. Oil Notches Record Growth," *Wall Street Journal*, June 13, 2013.

Jones, Charisse, "What's made from oil goes way beyond the gas pump," *USA Today*, Jan. 19, 2016.

Jones, Dan, *The Plantagenets: The Warrior Kings and Queens Who Made England*, Revised Edition, Viking, Penguin Group (USA), New York, 2013.

Joy, Bill, "Why the Future Doesn't Need Us," *Wired*, April 2000.

Kaczynski, Theodore John, *Technological Slavery*, Second Edition, Feral House, Port Townsend, Washington, 2010; Third Edition, Vol. 1, Fitch & Madison Publishers, Scottsdale, Arizona, 2019. Passages cited in the present work appear in both editions of *Technological Slavery*, except where otherwise noted.

Kaufmann, Walter (ed.), *The Portable Nietzsche*, Penguin Books, New York, 1976.

Kee, Robert, *The Green Flag: A History of Irish Nationalism*, Penguin Books, London, 2000.

Keefe, Patrick Radden, "Cat-and-Mouse Games," *New York Review*, May 26, 2005.

Keegan, John, *The Second World War*, Penguin Books, New York, 1990.

Keiper, Adam, "The Nanotechnology Revolution," *The New Atlantis: A Journal of Technology and Society*, Number 2, Summer 2003.

Kelly, Kevin, *What Technology Wants*, Penguin Books, New York, 2011.

Kendrick, Thomas Downing, *A History of the Vikings*, Dover Publications, Mineola, New York, 2004.

Kerr, Richard A., "Life Goes to Extremes in the Deep Earth—and Elsewhere?," *Science*, Vol. 276, No. 5313, May 2, 1997, pp. 703–04.

Kiviat, Barbara, "Below the Line," *Time*, Nov. 28, 2011.

Klein, Naomi, "Capitalism vs. the Climate," *The Nation*, Nov. 28, 2011.

Klemm, Friedrich, *A History of Western Technology*, trans. by Dorothea Waley Singer, M.I.T. Press, Cambridge, Massachusetts, 1964, sixth printing, 1978.

Knab, Sophie Hodorowicz, "Polish Farm Family Paid the Ultimate Price for Hiding Jews," *Polish American Journal*, Nov. 2012.

Koch, Wendy, "Nuclear industry sees new generation of reactors," *USA Today*, Nov. 27, 2012.

Kosthorst, Erich, *Die deutsche Opposition gegen Hitler zwischen Polen- und Frankreichfeldzug*, 3. bearbeitete Auflage, Schriftenreihe der Bundeszentrale für Heimatdienst, Heft 8, Bonn, 1957.

Krauss, Clifford, "South African Company to Build U.S. Plant to Convert Gas to Liquid Fuels," *New York Times Business*, Dec. 4, 2012.

Krishnamurthy, Nagaiyar, and Chiranjib Kumar Gupta, *Extractive Metallurgy of Rare Earths*, Second Edition, CRC Press, Taylor & Francis Group, Boca Raton, Florida, 2016.

Kunzig, Robert, "World Without Ice," *National Geographic*, Oct. 2011.

Kurzweil, Ray, *The Singularity is Near*, Penguin Books, New York, 2006.

La Botz, Dan, *Democracy in Mexico: Peasant Rebellion and Political Reform*, South End Press, Boston, 1995.

LeBlanc, Steven A., *Constant Battles: The Myth of the Peaceful, Noble Savage*, St. Martin's Press, New York, 2003.

Lee, Martha F., *Earth First!: Environmental Apocalypse*, Syracuse University Press, Syracuse, New York, 1995.

Leger, Donna Leinwand and Anna Arutunyan, "Meet the architects of data theft," *USA Today*, March 6, 2014.

Lenin, Vladimir Ilich, *Lenin on Organization*, Lenin Library, Daily Worker Publishing Company, Chicago, 1926.

Lenin, Vladimir Ilich, *Collected Works*, International Publishers, New York. We cite two different editions, that of 1929 and that of 1942.

Leslie, John, "Return of the killer nanobots," *Times Literary Supplement*, Aug. 1, 2003.

Leuchtenburg, William E., *Franklin D. Roosevelt and the New Deal, 1932–1940*, Harper & Row, New York, 1963.

Levin, Simon A. (ed.), *Princeton Guide to Ecology*, Princeton University Press, Princeton, New Jersey, 2009.

Lidtke, Vernon L., *The Outlawed Party: Social Democracy in Germany, 1878-1890*, Princeton University Press, Princeton, New Jersey, 1966.

Lieberman, Bruce, "Meanwhile, on a Planet Nearby...," *Air & Space*, June/July 2013.

Lindstrom, Martin, *Brandwashed: Tricks Companies Use to Manipulate Our Minds and Persuade Us to Buy*, First Edition, Crown Publishing Group, a division of Random House, New York, 2011.

Lipsher, Steve, "Guilty pleas unveil the tale of eco-arson on Vail summit," *The Denver Post*, Dec. 15, 2006.

Lockwood, Lee, *Castro's Cuba, Cuba's Fidel*, Macmillan, New York, 1967.
Lohr, Steve, "More Jobs Predicted for Machines, Not People," *New York Times*, Late Edition (East Coast), Oct. 24, 2011.
Lomborg, Bjorn, "Trump exposes climate deal," *USA Today*, March 30, 2017.
Lorenz, Edward N., *The Essence of Chaos*, University of Washington Press, Seattle, 1993.
Lovich, Jeffrey E., and Joshua R. Ennen, "Wildlife Conservation and Solar Energy Development in the Desert Southwest, United States," *Bioscience*, Vol. 61, No. 12, Dec. 2011, pp. 982–992.
Lukowski, Jerzy, and Hubert Zawadzki, *A Concise History of Poland*, Second Edition, Cambridge University Press, Cambridge, U.K., 2006, fifth printing, 2011.
Lynch, John, *Argentine Caudillo: Juan Manuel de Rosas*, SR Books, an imprint of Rowman & Littlefield Publishers, Lanham, Maryland, 2006.
MacFadyen, Dugald, *Alfred the West Saxon, King of the English*, J.M. Dent & Co., London, 1901.
MacLeod, Calum, "China sees unfulfilled potential in wind," *USA Today*, Sept. 28, 2009.
Malpass, Michael A., *Daily Life in the Inca Empire*, Greenwood Press, Westport, Connecticut, 1996.
Manchester, William, *The Arms of Krupp, 1587–1968*, Bantam Books, Toronto, 1970.
Manjoo, Farhad, *True Enough: Learning to Live in a Post-Fact Society*, John Wiley & Sons, Hoboken, New Jersey, 2008.
Mann, Charles C., "What if We Never Run Out of Oil?," *The Atlantic*, May 2013.
Mao Zedong (Tsetung), *Selected Readings from the Works of Mao Tsetung*, Foreign Languages Press, Peking (Beijing), 1971.
Margonelli, Lisa, "Down and Dirty," *The Atlantic*, May 2009.
Markoff, John, "Ay Robot! Scientists Worry Machines May Outsmart Man," *New York Times*, July 26, 2009.
Markoff, John, "Skilled Work, Without the Worker," *New York Times*, Aug. 19, 2012, Late Edition (East Coast).
Markoff, John, "Pentagon Offers a Robotics Prize," *New York Times*, Oct. 29, 2012, New York Edition.
Martin, Joseph Plumb, *Memoir of a Revolutionary Soldier*, Dover Publications, Mineola, New York, 2006.
Martínez Ruiz. See Azorín.
Marx, Karl, and Friedrich Engels, *The Communist Manifesto*, trans. by Samuel Moore, Simon & Schuster, New York, 1964.
Matheny, Keith, "Solar plans pit green vs. green: Renewable energy projects could threaten species, habitat," *USA Today*, June 2, 2011.
Matthews, Herbert L., *Fidel Castro*, Simon & Schuster, New York, 1969.

McCaffrey, Carmel, and Leo Eaton, *In Search of Ancient Ireland: The Origins of the Irish from Neolithic Times to the Coming of the English,* New Amsterdam Books, Ivan R. Dee, Publisher, Chicago, 2002.

McCullough, David, *John Adams*, Simon & Schuster, New York, 2002.

McKibben, Bill, *Enough: Staying Human in an Engineered Age*, Times Books, New York, 2003.

McKinney, Michael L., and Julie L. Lockwood, "Biotic homogenization: a few winners replacing many losers in the next mass extinction," *Trends in Ecology and Evolution*, Vol. 14, Issue 11, Nov. 1999, pp. 450–53.

Mellow, Craig, "Jet Race: In 1956, the Soviets held first place—briefly," *Air & Space*, Oct./Nov. 2013.

Milstein, Michael, "Pilot not included," *Air & Space*, June/July 2011.

Mote, Frederick W., *Imperial China, 900–1800*, Harvard University Press, Cambridge, Massachusetts, 2003.

Murphy, Audie, *To Hell and Back*, Owl Books, Henry Holt and Company, New York, 2002.

Naess, Arne, *Ecology, Community and Lifestyle: Outline of an Ecosophy*, trans. by David Rothenberg, Cambridge University Press, Cambridge, U.K., 1989.

Naruo Uehara, "Green Revolution" (editorial), *Japan Medical Association Journal*, Vol. 49, No. 7&8, July/Aug. 2006, p. 235.

Neusner, Jacob, and Bruce Chilton (eds.), *Altruism in World Religions*, Georgetown University Press, Washington, D.C., 2005.

Nevins, Allan, *Study in Power: John D. Rockefeller, Industrialist and Philanthropist* (2 Vols.), Charles Scribner's Sons, New York, 1953.

Nissani, Moti, *Lives in the Balance: The Cold War and American Politics, 1945–1991*, Hollowbrook Publishing/ Dowser Publishing Group, Wakefield, New Hampshire/ Carson City, Nevada, 1992.

Norris, Frank, *The Octopus: A Story of California*, Doubleday, Garden City, New York, 1947.

Noyes, J.H., *History of American Socialisms*, J.B. Lippincott & Co., Philadelphia, 1870.

Okada. See Yukinori.

Okrent, Daniel, *Last Call: The Rise and Fall of Prohibition*, Scribner, a division of Simon & Schuster, New York, 2010.

O'Regan, Davin, "Narco-states: Africa's next menace," *International Herald Tribune*, March 13, 2012.

Orr, H. Allen, "The God Project," *The New Yorker*, April 3, 2006.

Packer, George, "Knowing the Enemy," *The New Yorker*, Dec. 18, 2006.

Padgett, Tim, and Ioan Grillo, "Mexico's Meth Warriors," *Time*, June 28, 2010.

Pandita, Rahul, *Hello, Bastar: The Untold Story of India's Maoist Movement*, Tranquebar Press, Chennai, 2011.

Parker, Geoffrey (ed.), *The Cambridge History of Warfare*, Cambridge University Press, Cambridge, U.K., 2008.

Patterson, James T., *America in the Twentieth Century: A History*, Fifth Edition, Harcourt College Publishers, Fort Worth, Texas, 2000.

Payne, Stanley G., *El Franquismo, Segunda Parte, 1950–1959. Apertura exterior y planes de estabilización*, Arlanza Ediciones, Madrid, 2005.

Paz, Fernando, *Europa bajo los escombros: Los bombardeos aéreos en la Segunda Guerra Mundial*, Áltera, Barcelona, 2008.

Peck, Don, "They're Watching You at Work," *The Atlantic*, Dec. 2013.

Perrow, Charles, *Normal Accidents: Living with High-Risk Technologies*, Basic Books, New York, 1984.

Perrow, Charles, *The Next Catastrophe: Reducing our vulnerabilities to natural, industrial, and terrorist disasters*, Princeton University Press, Princeton, New Jersey, 2007.

Peterson, Christopher L., "Fannie Mae, Freddie Mac, and the Home Mortgage Foreclosure Crisis," *Loyola University New Orleans Journal of Public Interest Law*, Vol. 10, 2009, pp. 149–170.

Pipes, Richard (ed.), *The Unknown Lenin: From the Secret Archive*, Yale University Press, New Haven, Connecticut, 1998.

Pirenne, Henri, *Mohammed and Charlemagne*, Meridian Books, The World Publishing Company, Cleveland, Ohio, 1957.

Plato. See Buchanan.

Quammen, David, "How Animals and Humans Exchange Disease," *National Geographic*, Oct. 2007.

Radzinsky, Edvard, *The Last Tsar: The Life and Death of Nicholas II*, trans. by Marian Schwartz, Doubleday, New York, 1992.

Randall, Willard Sterne, *Thomas Jefferson: A Life*, Harper Collins, New York, 1994.

Read, Anthony, and David Fisher, *The Fall of Berlin*, Da Capo Press, New York, 1995.

Read, Piers Paul, *The Templars*, Da Capo Press, New York, 2001.

Reed, Stanley, "Shell Bets on a Colossal Floating Liquefied Natural Gas Factory Off Australia," *New York Times Business*, Nov. 13, 2012.

Reid, P.R., *The Colditz Story*, J.B. Lippincott Company, Philadelphia, 1953.

Reid, Stuart A., "'Let's Go Take Back Our Country.'" *The Atlantic*, March 2016.

Remnick, David, "The Talk of the Town," *The New Yorker*, Aug. 25, 2008.

Ribas, Ignasi, "The Sun and stars as the primary energy input in planetary atmospheres," published by the International Astronomical Union as *Solar and Stellar Variability: Impact on Earth and Planets, Proceedings IAU Symposium, No. 264, 2009*.

Rifkin, Jeremy, *The Third Industrial Revolution: How Lateral Power is Transforming Energy, the Economy, and the World*, Palgrave Macmillan, a division of St. Martin's Press, New York, 2011.

Ripley, Amanda, "To Catch a Drone," *The Atlantic*, Nov. 2015.

Rosenthal, Elizabeth, "U.S. Is Forecast to be No. 1 Oil Producer," *New York Times Business*, Nov. 13, 2012.

Rossi, A., *The Rise of Italian Fascism, 1918–1922*, trans. by Peter and Dorothy Wait, Methuen, London, 1938.

Rossiter, Clinton, *The American Presidency,* Time, Inc. Book Division, New York, 1960.

Rothfels, Hans, *Deutsche Opposition gegen Hitler: Eine Würdigung*, Neue, erweiterte Ausgabe, Fischer Taschenbuch Verlag, Frankfurt am Main, 1986.

Rothkopf, David, "Command and Control," *Time*, Jan. 30, 2012.

Rotman, David, "How Technology is Destroying Jobs," *MIT Technology Review*, Vol. 116, No. 4, July/Aug. 2013, pp. 28–35.

Ruiz. See Azorín.

Runciman, Steven, *A History of the Crusades*, Vol. III, *The Kingdom of Acre and the Later Crusades*, The Folio Society, London, fifth printing, 1996.

Russell, Diana E.H., *Rebellion, Revolution, and Armed Force: A Comparative Study of Fifteen Countries with Special Emphasis on Cuba and South Africa*, Academic Press, New York, 1974.

Sabato, Hilda, and Alberto Lettieri (eds.), *La vida política en la Argentina del siglo XIX: Armas, votos y voces*, Primera edición, Fondo de Cultura Económica de Argentina, Buenos Aires, 2003.

Sallust (Gaius Sallustius Crispus), *The Jugurthine War; The Conspiracy of Catiline,* trans. by S.A. Handford, Penguin Books, Baltimore, 1967.

Sampson, Anthony, *Mandela: The Authorized Biography*, Alfred A. Knopf, New York, 1999.

Saney, Isaac, *Cuba: A Revolution in Motion,* Fernwood Publishing, London, 2004.

Saporito, Bill, "Hack Attack," *Time*, July 4, 2011.

Sarewitz, Daniel, and Roger Pielke, Jr., "Learning to Live With Fossil Fuels," *The Atlantic*, May 2013.

Sarmiento, Domingo Faustino, *Facundo. Civilización y Barbarie*, edited by Roberto Yahni, Séptima edición, Ediciones Cátedra (Grupo Anaya, S.A.), Madrid, 2005.

Schebesta, Paul, *Die Bambuti-Pygmäen vom Ituri*, II. Band, I. Teil, Institut Royal Colonial Belge, Brussels, 1941.

Searcy, Dionne, "Trying to 'Salvage' Nation, Nigeria's President Faces a Crisis in Every Direction," *New York Times International*, July 18, 2016.

Seligman, Martin E.P., *Helplessness: On Depression, Development, and Death*, W.H. Freeman and Company, New York, 1992.

Selznick, Philip, *The Organizational Weapon: A Study of Bolshevik Strategy and Tactics*, The Free Press of Glencoe, Illinois, 1960.

Shapiro, Fred R. (ed.), *The Yale Book of Quotations*, Yale University Press, New Haven, Connecticut, 2006.

Sharer, Robert J., *The Ancient Maya*, Fifth Edition, Stanford University Press, Stanford, California, 1994.

Sharkey, Patrick, "The Eviction Curse," *The Atlantic*, June 2016.
Shattuck, Roger, "In the Thick of Things," *The New York Review*, May 26, 2005.
Shukman, Henry, "After the apocalypse," *The Week*, April 1, 2011.
Silverman, Kenneth (ed.), *Benjamin Franklin: The Autobiography and Other Writings*, Penguin Books, New York, 1986.
Skidelsky, Robert, *John Maynard Keynes*, Vol. Three, *Fighting for Freedom, 1937–1946*, Viking Penguin, New York, 2001.
Skrbina, David (ed.), *Confronting Technology*, Creative Fire Press, Detroit, 2010.
Smelser, Neil J., *Theory of Collective Behavior*, Macmillan, New York, 1971.
Smith, Alice Kimball, and Charles Weiner (eds.), *Robert Oppenheimer: Letters and Recollections,* Stanford University Press, Stanford, California, 1995.
Sodhi, Navjot S., Barry W. Brook, and Corey J.A. Bradshaw, "Causes and Consequences of Species Extinctions," in Levin, pp. 514–520.
Sohail Ejaz. See Ejaz.
Somers, James, "The Man Who Would Teach Machines to Think," *The Atlantic*, Nov. 2013.
Soriano, Graciela (ed.), *Simón Bolívar: Escritos políticos,* Alianza Editorial, Madrid, 1975.
Stafford, David, *Secret Agent: The True Story of the Covert War Against Hitler*, The Overlook Press, New York, 2001.
Stalin, J., *Foundations of Leninism*, International Publishers, New York, 1932.
Stalin, J., *History of the Communist Party of the Soviet Union (Bolsheviks): Short Course*, Prism Key Press, New York, 2013. Though the authorship was attributed to Stalin, this book was mostly written by a commission of the Central Committee of the Communist Party of the Soviet Union. Selznick, p. 42n23. Ulam, p. 638. The copy referenced here was manufactured in Lexington, Kentucky on July 29, 2014, presumably through some sort of print-on-demand system.
Starr, Chester G., *The Origins of Greek Civilization, 1100–650 B.C.*, W.W. Norton & Company, New York, 1991.
Steele, David Ramsay, *From Marx to Mises: Post-Capitalist Society and the Challenge of Economic Calculation,* Open Court, La Salle, Illinois, 1992.
Stigler, George J., *The Theory of Price,* Fourth Edition, Macmillan, New York, 1987.
Suárez, Luis, *Franco: Crónica de un tiempo. Victoria frente al bloqueo. Desde 1945 hasta 1953,* Editorial Actas, Madrid, 2001.
Sueiro, Daniel, and Bernardo Díaz Nosty, *Historia del Franquismo,* Vol. I, Sociedad Anónima de Revistas, Periódicos y Ediciones, Madrid, 1986.
Surowiecki, James, "Fuel for thought," *The New Yorker,* July 23, 2007.
Swift, Jonathan. See Browning.
Taagepera, Rein, "Size and Duration of Empires: Systematics of Size," *Social Science Research*, Vol. 7, 1978, pp. 108–127.

Tacitus, Publius (or Gaius) Cornelius, *The Annals*, trans. by Alfred John Church and William Jackson Brodribb, Dover Publications, Mineola, New York, 2006.

Tannenbaum, Frank, *Peace by Revolution: Mexico After 1910*, Columbia University Press, New York, 1966.

Taylor, Neill, et al., "Resolving safety issues for a demonstration fusion power plant," *Fusion Engineering and Design*, Vol. 124, Nov. 2017, pp. 1177–1180. Apparently this journal as a whole is not available to the public, but some articles, including the one cited here, are available on the Internet at https://www.sciencedirect.com/science/article/pii/S0920379617301011. The selfsame copy used by this writer should be available in the University of Michigan's Special Collections Library at Ann Arbor.

Tella, Torcuato S. di, Gino Germani, Jorge Graciarena y colaboradores, *Argentina, Sociedad de Masas*, Tercera edición, Editorial Universitaria de Buenos Aires, 1971.

Ternavasio, Marcela, "La visibilidad del consenso. Representaciones en torno al sufragio en la primera mitad del siglo XIX," in Sabato & Lettieri, pp. 57–73.

Thurston, Robert W., *Life and Terror in Stalin's Russia, 1934–1941*, Yale University Press, New Haven, Connecticut, 1996.

Tipton, Frank B., "The National Consensus in German Economic History," *Central European History*, Vol. 7, No. 3 (Sept. 1974), pp. 195–224.

Tocqueville, Alexis de, *Democracy in America* (2 Vols.), Vintage Books, Random House, New York, 1945.

Trees, Andrew, "Founders would dump Trump," *USA Today*, March 31, 2016.

Trotsky, Leon, *History of the Russian Revolution*, trans. by Max Eastman, Pathfinder, New York, 1980.

Turnbull, Colin M., *The Forest People*, Simon and Schuster, New York, 1962.

Turnbull, Colin M., *Wayward Servants: The Two Worlds of the African Pygmies*, Natural History Press, Garden City, New York, 1965.

Turnbull, Colin M., *The Mbuti Pygmies: Change and Adaptation*, Harcourt Brace College Publishers, Fort Worth, Texas, 1983.

Uehara. See Naruo.

Ulam, Adam B., *Stalin: The Man and His Era*, Beacon Press, Boston, 1987.

Utt, Ronald D., "The Subprime Mortgage Market Collapse: A Primer on the Causes and Possible Solutions," *Backgrounder* No. 2127, April 22, 2008 (published by The Heritage Foundation).

Vance, Ashlee, "Merely Human? So Yesterday," *New York Times Sunday Business*, June 13, 2010.

Vara, Vauhini, "How Frackers Beat OPEC," *The Atlantic*, Jan./Feb. 2017.

Vassilyev, A.T., *The Ochrana*, J.B. Lippincott Company, Philadelphia, 1930.

Vergano, Dan, "Mobster myths revisited," *USA Today*, June 20, 2013.

Vick, Karl, "The Ultra-Holy City," *Time*, Aug. 13, 2012.

Von Harnack. See Harnack.

Wald, Matthew L., "Nuclear Industry Seeks Interim Site to Receive Waste," *New York Times*, Aug. 27, 1993.

Wald, Matthew L., "What Now for Nuclear Waste?," *Scientific American*, Vol. 301, No. 2, Aug. 2009, pp. 46–53.

Walsh, Bryan, "The Gas Dilemma," *Time*, April 11, 2011.

Walsh, Bryan, "Power Surge: The U.S. is undergoing an energy revolution," *Time*, Oct. 7, 2013.

Walston, Leroy J., Jr., et al., "A preliminary assessment of avian mortality at utility-scale solar energy facilities in the United States," *Renewable Energy*, Vol. 92, July 2016, pp. 405–414.

Walter, Eugene Victor, *Terror and Resistance: A Study of Political Violence*, Paperback Edition, Oxford University Press, New York, 1972.

Ward, Peter, *The Medea Hypothesis*, Princeton University Press, Princeton, New Jersey, 2009.

Watson, Traci, "Gold Rush days leave toxic legacy," *USA Today*, Oct. 29, 2013.

Watts, Meriel, *Pesticides: Sowing Poison, Growing Hunger, Reaping Sorrow*, Second Edition, Pesticide Action Network Asia and the Pacific, 2010.

Weber, Max, *Die protestantische Ethik und der Geist des Kapitalismus*, Verlag von J.C.B. Mohr (Paul Siebeck), Tübingen, 1934. This is an offprint of pages 1–206 of Weber's *Gesammelte Aufsätze zur Religionssoziologie*, I. Band, Dritte photomechanisch gedruckte Auflage, 1922, of the same publisher.

Weir, Alison, *Henry VIII: The King and His Court*, Ballantine Books, New York, 2008.

Weise, Elizabeth, "DIY 'biopunks' want science in hands of people," *USA Today*, June 1, 2011.

Weise, Elizabeth, "Invasive species blighting the landscape," *USA Today*, Nov. 28, 2011.

Welch, William M., "Bird deaths soar at wind farms," *USA Today*, Sept. 22, 2009.

Whittle, Richard, "The Drone Started Here," *Air & Space*, April/May 2013.

Wilkinson, Todd, *Science Under Siege: The Politicians' War on Nature and Truth*, Johnson Books, Boulder, Colorado, 1998.

Wissler, Clark, *Indians of the United States*, Revised Edition, Anchor Books, Random House, New York, 1989.

Wolk, Martin, "Combat Missions of American Airmen to Poland in World War II: A Story of Charles Keutman," *Polish American Journal*, June 2012.

Woo, Elaine, "Pole saved 2,500 Jewish kids in WWII," *The Denver Post*, May 13, 2008.

Wood, Graeme, "Moving Heaven and Earth," *The Atlantic*, July/Aug. 2009.

Woodall, Curt, Letter to editor, *Air & Space*, Feb. 2011.

Wu, Tim, *The Attention Merchants: The Epic Scramble to Get Inside Our Heads*, First Vintage Books Edition, Penguin Random House, New York, 2017.

Yukinori Okada and Susumu Wakai, "The Longitudinal Effects of the 'Green Revolution' on the Infant Mortality Rate in Thailand," *Japan Medical Association Journal*, Vol. 49, No. 7 & 8, July/Aug. 2006, pp. 236–242.

Zakaria, Fareed, "Don't Make Hollow Threats," *Newsweek*, Aug. 22, 2005.

Zakaria, Rafiq, *The Struggle Within Islam*, Penguin Books, London, 1989.

Zierenberg, Robert A., Michael W.W. Adams, and Alissa J. Arp, "Life in extreme environments: Hydrothermal vents," *Proceedings of the National Academy of Sciences U.S.A.*, Vol. 97, No. 24, Nov. 21, 2000, pp. 12961–62.

Zimmermann, G.A., *Das Neunzehnte Jahrhundert: Geschichtlicher und Kulturhistorischer Rückblick,* Zweite Hälfte, Zweiter Theil, Druck und Verlag von Geo. Brumder, Milwaukee, 1902.

WORKS WITHOUT NAMED AUTHOR

Air & Space magazine

Anarchy: A Journal of Desire Armed

Atlantic, The

Bible, The Holy: King James Version; New English Bible; Revised English Bible; New International Version

Constitution of the United States

Denver Post, The

Economist, The

Encyclopedia of American Studies, published under the auspices of the American Studies Association by Grolier Educational, a division of Scholastic Incorporated, New York, 2001.

Evolutionary and Revolutionary Technologies for Mining, National Academy Press, Washington, D.C., 2002.

Federal Reporter

GMO Quarterly Letter, published by GMO Corporation

Green Anarchy newspaper

Historical Materialism (Marx, Engels, Lenin), Progress Publishers, 1972.

ISAIF = *Industrial Society and Its Future*, in Kaczynski.

Los Angeles Times, The

McGraw-Hill Encyclopedia of Science & Technology, Eleventh Edition, McGraw-Hill, New York, 2012.

National Geographic magazine

NEB = *The New Encyclopaedia Britannica*, Fifteenth Edition. The Fifteenth Edition has been modified every few years. We put a date in parentheses after NEB —e.g., NEB (2003)—to indicate the particular version of NEB that we cite.

New York Times, The

Newsweek

Polish American Journal
Popular Science
Science News
Scientific American
Select Committee to Study Governmental Operations With Respect to Intelligence Activities, Final Report, S. Rep. No. 755, Book II (Intelligence Activities and the Rights of Americans), and Book III (Supplementary Detailed Staff Reports on Intelligence Activities and the Rights of Americans), 94th Congress, Second Session (1976).
Time magazine
USA Today
US News & World Report
Vegetarian Times
Wall Street Journal, The
Warrior Wind No. 2, available in the Labadie Collection at the University of Michigan's Special Collections Library in Ann Arbor.
Week, The
Wired magazine
World Book Encyclopedia, The, editions of 2011 and 2015.

INDEX